CULTIVATING DIVERSITY

AGROBIODIVERSITY AND FOOD SECURITY

LORI ANN THRUPP

WORLD RESOURCES INSTITUTE
WASHINGTON, DC

CAROL ROSEN
PUBLICATIONS DIRECTOR

DEBORAH FARMER
MANAGING EDITOR

HYACINTH BILLINGS
PRODUCTION MANAGER

DEAN CHAPMAN/PANOS PICTURES
COVER PHOTO

Each World Resources Institute Report represents a timely, scholarly treatment of a subject of public concern. WRI takes responsibility for choosing the study topics and guaranteeing its authors and researchers freedom of inquiry. It also solicits and responds to the guidance of advisory panels and expert reviewers. Unless otherwise stated, however, all the interpretation and findings set forth in WRI publications are those of the authors.

CONTENTS

FOREWORD

People around the world are now recognizing that the diversity of plant and animal species is a shrinking treasure, no less for human endeavors than for natural ecosystems. This report, growing out of the World Resources Institute's research on sustaining biological resources, focuses on the impact of declining biodiversity in agriculture. It documents the threats to environmental resources and food supplies that result from the loss of agrobiodiversity.

Those threats are tied to the disappearance of crop and livestock species, the erosion of diversity among soil organisms and insects, and the narrowing of ecosystems. Each of these developments reduces productivity and jeopardizes food security worldwide. They therefore threaten basic livelihoods and economic development, as well as access to and control of germplasm, the basic genetic resources of plants and animals that are the building blocks of food and life.

Drawing upon experiences from many parts of the world, this report summarizes the vital role of agrobiodiversity for food security, productivity, and ecological sustainability. It analyzes the main causes of the decline in agrobiodiversity and recommends urgently needed policies and practices to integrate biodiversity into sustainable agriculture. The Convention on Biological Diversity has established mandates to address the loss of agrobiodiversity, but much remains to be done. Only if policymakers, producers, entrepreneurs, researchers, and even consumers recognize each other's concerns and act together will it be possible to implement the Convention's provisions and confront the threats to agrobiodiversity.

The guidelines proposed in this report would help meet the worldwide need for food, while protecting opportunities for farmers and communities. Our hope is to instigate a needed dialogue in implementing guidelines for sustaining agrobiodiversity.

JONATHAN LASH
PRESIDENT
WORLD RESOURCES INSTITUTE

ACKNOWLEDGMENTS

I wish to thank Nabiha Megateli, Rita Banerji, and Dina Matthews, research assistants for the project, who carried out an extensive literature review, helped to draft some of the text boxes, and provided other valuable research support. I also thank my WRI colleagues Kenton Miller, Walter Reid, Thomas Fox, and Arthur Getz for their valuable advice, reviews, and interest; Bob Blake, Paul Faeth, and Peter Veit for their comments on drafts; and Roberto Colque and Consuelo Holguin, who skillfully prepared graphics and provided assistance in production. I appreciate the advice of Miguel Altieri, Wanda Collins, Daniel Debouck, K.L. Heong, Calestous Juma, Patrick Madden, Jeffrey McNeely, John Mugabe, Patrick Mulvany, Christina Reinhard, Kristin Schafer, and David Williams, who reviewed draft versions of this report.

This report and WRI's project on agrobiodiversity have also benefitted from collaboration and dialogue among colleagues in Africa, Latin America, and Asia. WRI is grateful for the collaboration of the African Centre for Technology Studies (ACTS) and other organizations in East Africa for addressing this issue at the regional level and undertaking useful case studies from that region. This collaboration also generated a successful regional forum on agrobiodiversity issues held in November 1997 in Kenya that was jointly organized by ACTS, WRI, the International Centre for Research on Agroforestry, the World Conservation Union–East Africa Regional Office, and the World Wildlife Fund in East Africa.

We appreciate the support of the Swedish International Development Cooperation Agency (SIDA), which has generously funded WRI's project on agrobiodiversity. We also are thankful for the support of the Dutch Ministry of Foreign Affairs and the United States Agency for International Development (particularly Walter Knausenberger). We appreciate the contribution and reviews of the World Bank Environment and Agriculture Departments, specifically Jitendra Srivastava, Lars Vidaeus, Kathy Mackinnon, John Kellenberg, and Stefano Pagiola, and the Global Environment Facility, particularly Walter Lusigi, for working with WRI in this field and for reviews of earlier drafts. Last, but certainly not least, I owe thanks to rural men and women, with whom we have worked over many years, whose experiences in this area provide inspiration for work on sustainable agriculture. We look forward to receiving feedback on the report, and beyond this, to continue developing insights and actions that will contribute to agrobiodiversity conservation and enhancement.

L.A.T.

EXECUTIVE SUMMARY

THE AGRICULTURE-BIODIVERSITY NEXUS

There is a growing realization that biodiversity is a fundamental basis of agricultural production and food security, as well as a valuable ingredient of ecological stability. However, biodiversity associated with agriculture and food production (i.e., *agrobiodiversity*) is rapidly disappearing throughout the world. The loss of diversity extends from genetic resources in plants and animals to species diversity among crops, livestock, insects, and microorganisms. It even includes the narrowing of agroecosystems in general. A related and equally alarming situation is the loss of biodiversity in "natural" habitats caused by agricultural expansion into frontier areas. Such losses jeopardize productivity, threaten food security, and result in high economic as well as social costs. In many areas, the livelihoods and survival of local people are imperiled.

Agricultural development and biodiversity conservation are sometimes perceived as opposing interests. But in many cases, such conflicts do not exist and they are certainly not inevitable. In fact, evidence shows that integrating biodiversity and agriculture is beneficial for food production, ecosystem health, and for economically and ecologically sustainable growth. While agrobiodiversity has gained the attention of environment and development institutions in recent years, it is certainly not new. Ancient agricultural settlements made use of a variety of plants, livestock, and agroecosystems, and agrobiodiversity continues to be a fundamental feature of farming systems around the world. Agrobiodiversity conservation is also tied to the rich cultural diversity and local knowledge of women and men, with many principles from traditional systems still relevant today for large as well as small-scale production.

Recognizing the urgency of biodiversity losses, the Convention on Biological Diversity (CBD) has mandated actions that nations and institutions must implement to conserve agricultural biodiversity. Furthermore, the World Food Summit and other international conventions have established global mandates to fulfill the goal of "food for all." And yet, written agreements are insufficient to address these problems. What is urgently needed is effective action that can overcome conflicts and change conventional agricultural practices and economic policies.

THREATS TO AGROBIODIVERSITY

Despite the importance of agricultural biodiversity, the diversity of crop and livestock species currently in use worldwide is rapidly dwindling. Although people consume approximately 7,000 species of plants, only 150 species are commercially important. Just over 100 species account for 90 percent of the world's food crops. In fact, rice, wheat, and maize alone account for nearly 60 percent of the calories and 56 percent of the protein people derive from plants. There are also fewer types of farming systems and general cropping patterns, which raises the risks for farmers and leaves them more vulnerable to changes in climate. This narrowing of food crops ultimately undermines the stability of the food supply.

A loss of diversity in farms also leaves crops more vulnerable to pests and disease. Serious economic loss and human suffering are inevitable when monocultural, uniform varieties are attacked by pests. In addition, there has been a serious decline in soil organisms, which are vital for soil fertility and structure, and in beneficial insects and fungi. Such losses, along with reduced diversity of farming system types, further increase risks and reduce productivity.

The underlying causes of this loss of diversity consist of a complex mix of policies, practices, and pressures for economic and agricultural growth, as well as demographic changes and inequities in the control of resources. Although the policies of the Green Revolution, promoting monocultural systems, uniform crop varieties, and agrochemical inputs, did contribute to aggregate increases in production in many areas, these patterns have also eroded agricultural biodiversity and degraded other natural resources. At the same time, more than 800 million people suffer from hunger and malnutrition globally, and resources and food are distributed highly inequitably. These dilemmas pose tremendous challenges to meet growing food needs while conserving resources.

IMPERATIVES AND OPPORTUNITIES FOR CHANGE

A variety of groups and institutions (i.e., *stakeholders*) in many countries have attempted to address this predicament. At the grassroots level, growing numbers of farmers, rural community associations, nongovernmental organizations, consumers, and other members of civil society are increasingly involved in conserving agricultural biodiversity and in promoting the equitable distribution and right of access to genetic resources.

At the same time, international actors and agreements have provided a framework for change. In addition to the Convention on Biological Diversity (CBD) mandate, other international conventions affecting trade and intellectual property rights also affect access to plant genetic resources for public institutions and private companies as well as farmers. The Trade-Related Intellectual Property Rights (TRIPS) section of the General Agreement on Tariffs and Trade (GATT) influences the control, sale, and access to genetic resources and property rights tied to agrobiodiversity. Yet, such agreements and the control by the World Trade Organization in this context present major dilemmas because they conflict with the CBD, establish private regimes over intellectual property, and do not adequately value local peoples' rights.

Significant international institutions influencing the use of genetic resources include the Consultative Group for International Agricultural Research, the International Plant Genetic Resources Institute, the Food and Agriculture Organization, and the Commission on Plant Genetic Resources. Some government agencies and research institutions have paid attention to agrobiodiversity issues, for example, by developing gene banks for conservation of plant genetic resources, and/or undertaking research programs. Other agencies have begun extension or development projects that potentially contribute to the mandates from the CBD.

Yet, additional work and policy changes are urgently needed to implement the agreements of the Convention, to strengthen, expand, and coordinate the initiatives for agrobiodiversity conservation and enhancement, and to overcome the real challenges and barriers that perpetuate these losses.

The strategic principles shown below are needed at all levels and have proven effective as general approaches to protect and enhance agrobiodiversity. If these changes and principles are *not* implemented, all of humanity will be threatened by increasing food insecurity.

In practical terms, at the local level, these principles translate into the urgent need for integrated pest, crop, and soil management methods, organic and regenerative approaches, and changes in agricultural research paradigms. Success hinges on the empowerment of local people in research, development, and decisionmaking.

Equally important at a macro level is the imperative for major reforms in agricultural and economic policies and institutions to ensure that there are effective capacities and political

MAJOR RECOMMENDATIONS

PRINCIPLES AND POLICIES TO ENHANCE AGROBIODIVERSITY

The following principles are recommended strategic guidelines:

→ Support sustainable ecological agriculture, which includes the goals of food security, social equity and health, economic productivity, and ecological integrity, as a framework for enhancing agrobiodiversity.

→ Develop an ecosystems approach, using agroecology as a guiding scientific paradigm, to support and validate the sustainable use and enhancement of agrobiodiversity at all levels.

→ Empower farmers and communities to protect their rights to resources, support their knowledge and cultural diversity, and ensure their participation in decisionmaking and conservation.

→ Adapt agricultural practices and land use to local agroecological and socioeconomic conditions, adjusted to local diverse needs and aspirations, and building upon local successful experiences.

→ Conserve and regenerate plant and animal genetic resources and ecosystem services using agroecological and socially beneficial methods for sustainable intensification and biodiversity enhancement.

→ Develop policies and institutional changes that support agrobiodiversity, ensure food security, and protect farmers' rights and eliminate policies that promote uniform monocultural systems.

→ Uphold and implement agrobiodiversity provisions of the Convention on Biological Diversity, as well as the mandates of the World Food Summit.

support to implement agrobiodiversity conservation strategies. This depends on far-reaching political commitments by governments. Without significant policy transformations, it is unlikely that agrobiodiversity-enhancing practices can be widely adopted. Among the most crucial changes are the elimination of policies that erode agrobiodiversity (such as subsidies and incentives for agrochemicals and high-yield varieties) and the adoption of market and trade policies that incorporate ecological concerns. It is also essential to implement laws and other measures to ensure ethical business practices by agricultural technology companies and to prevent their unfair control over plant genetic resources.

High economic returns and production increases have been reported by producers and groups who are incorporating such principles and practices. These approaches can also meet food needs without extensification, thereby reducing pressure on biodiversity, both on and off farms. The types of practices and policies outlined in the report constitute promising "win-win" opportunities to merge the goals of food production and agrobiodiversity conservation.

1

INTRODUCTION AND
FRAMEWORK

"Agricultural biodiversity is a matter of life and death for us.... We cannot separate agrobiodiversity from food security."

—Zambian delegate to the Conference of Parties, Convention on Biological Diversity, May 1998

There is a growing realization that bio-diversity is a fundamental basis of agricultural production and food security, as well as a valuable ingredient of ecological stability. Agricultural biodiversity, or "agrobiodiversity," has been called the cornerstone of stability—a basis of livelihoods and of sustainable development.[1] The term encompasses not only diversity among plant and animal genetic resources, soil organisms, insects, and other flora and fauna in managed ecosystems (agro-ecosystems), but also diversity among elements of natural habitats that pertain to food production. Agrobiodiversity makes it possible for farmers to recycle nutrients, reduce pest and disease problems, control weeds, maintain good soil and water conditions, and handle climatic stress, while producing agricultural products necessary for health and human survival. It therefore has multiple economic, ecological, and social benefits.

At the same time, however, biodiversity is being seriously eroded by agricultural development in many areas. Unsustainable patterns of agricultural production undermine and conflict with biodiversity. The decline of diversity in genetic resources, crop varieties, insects, soil and aquatic organisms, agroecosystems, and the "wild" resources that surround farmlands undermines productivity and food security and leads to irreversible biological losses that have a high socio-economic price.

Agricultural growth and biodiversity conservation are not always conflicting goals, however. There is ample evidence, both past and present, to show the multiple benefits of integrating biodiversity and agriculture for both small- and large-scale farming.[2] The conflicts can be overcome, and this integration can and must be achieved through the use of sustainable ecological practices and major changes in policies, institutions, and paradigms.

Agricultural biodiversity lies within the general framework of sustainable human development, which envisions that critical needs for food security, economic productivity, social equity and health, and ecological integrity can be simultaneously achieved. *(See Figure 1.)* Food security is a particularly important element and is considered a basic human right in this analysis. It means, in general terms, "access to food for a healthy life by all people at all times."[3]

This report reveals the urgent need for action to reverse the present trend, provides justification for mainstreaming biodiversity in agricultural development, and highlights effective prac-

tices and policies for this purpose. It begins by clarifying basic concepts and benefits of agrobiodiversity, stressing not only the value of genetic resources conservation, but also agroecosystem approaches. It then summarizes the problems of agrobiodiversity loss and its underlying causes. Finally, the report identifies policies and practices that can be "win-win" solutions including practical ways to merge agriculture, food security, and biodiversity conservation.

Several major international bodies have recently focused on the issue of agricultural biodiversity. The Global Convention on Biological Diversity (CBD) highlights agrobiodiversity as a key concern, mandating action and change. *(See Box 1.)* This important international convention requires that signatory nations implement specific actions and policies to ensure sustainable use and enhancement of agrobiodiversity. The 1996 World Food Summit also put forth an international mandate to achieve the goal of "food for all," and overcome the problem of hunger affecting some 800 million people.[4] While these political affirmations are important, they are not sufficient to bring about change. Concerted actions are needed at all levels to merge agriculture and environmental interests.

The information in this report is intended for planners, policymakers, producers, NGOs, researchers, businesses, and the broader public. It stresses ways to strengthen grassroots innovations and farmers' experiences in using agrobiodiversity. There are also suggested guidelines and strategies to implement the CBD and address the related problems of agrobiodiversity conservation and food security. This paper shows how agricultural and environmental interests can and must merge. If this does not happen, the world's valuable resources and food supplies will be further jeopardized.

FIGURE 1.

INTEGRATING AGRICULTURE AND BIODIVERSITY TO ACHIEVE SUSTAINABLE DEVELOPMENT

ECOLOGICAL
- Environmental soundness
- Ecological health/integrity
- Natural resource management

SUSTAINABLE DEVELOPMENT

ECONOMIC
- Food security
- Economic viability
- Agricultural productivity
- Policy support

SOCIAL
- Empower rural poor
- Social equity
- Healthy and safe for people
- Public participation

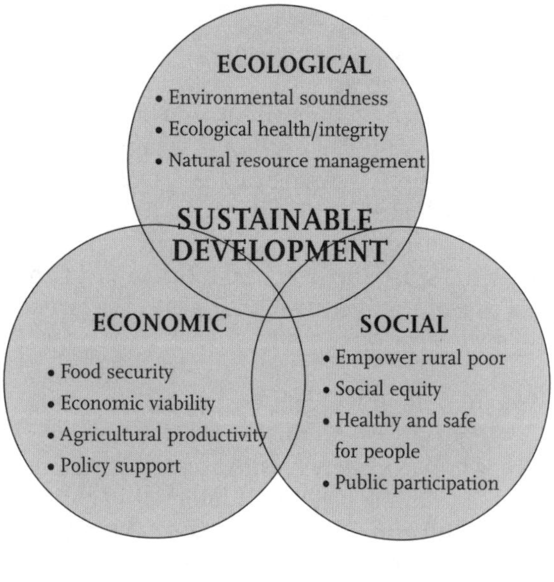

Source: L. A. Thrupp, World Resources Institute.

BOX 1.

Recommendations Summarized from the Global Convention on Biodiversity

- Identify key components of biological diversity in agricultural production systems.
- Redirect support measures that counter the objectives of the Convention on Biological Diversity.
- Internalize environmental costs in assessments of agriculture and economic development.
- Implement targeted incentive measures that have positive impacts on agrobiodiversity while also enhancing sustainable agriculture.
- Encourage the development of technologies and practices that increase productivity and also arrest degradation, and reclaim and restore biological diversity.
- Empower indigenous and local communities and build capacity for *in situ* conservation and sustainable use of agricultural biodiversity.
- Encourage evaluation of impacts on biodiversity from agricultural development projects.
- Link agrobiodiversity conservation efforts to programs affecting marine, coastal, and freshwater ecosystems.
- Promote partnerships with researchers, extension workers, and farmers for agrobiodiversity.
- Promote appropriate research and services for farmers that are based on genuine partnerships.
- Promote research and development on integrated pest management, particularly methods that maintain biodiversity.

- Encourage regulations and/or measures to ensure appropriate use of and to discourage excessive dependence on agrochemicals with a view to reducing impacts on biodiversity.
- Study and use methods and indicators to monitor impacts of agricultural projects on agrobiodiversity.
- Study the positive and negative impacts of agricultural intensification and extensification on ecosystems and biomes.

Recommendations Summarized from the World Food Summit

- Create a social and economic environment to eradicate hunger and support a durable peace.
- Eradicate poverty and inequality, improving physical and economic access by all.
- Develop participatory sustainable agriculture, fisheries, and forestry through intensified and diversified food systems and technologies and integrated approaches to food security.
- Ensure food and agricultural trade that is conducive to food security through fair trade.
- Prevent and be prepared for natural disasters and manmade emergencies.
- Ensure optimal allocation of public and private investments for food security.

Source: These statements are paraphrased based on the Conference of Parties. UNEP/CBD/COP/3/L12 and the World Summit Plan of Action for World Food Security.

2

BENEFITS OF AGROBIODIVERSITY

"Agricultural biodiversity is a basis of... sustainable development... Our survival depends on equitable access, sharing of plant genetic resources, and respecting rights of farmers...."

—Tanzanian delegate to the Conference of Parties, Convention on Biological Diversity, May 1998

BIODIVERSITY IN FOOD PRODUCTION

The concept of agricultural biodiversity has been popularized in recent years by environment and development institutions; yet, it is by no means new. Biodiversity and peoples' intimate knowledge about it have been fundamental to food provision since agriculture was invented some 12,000 years ago.[5] While farmers, fishers, and herders rarely use this technical term, they continually manage and use biodiversity, and appreciate nature's services and advantages for agriculture. Likewise, food production is actually dependent on a rich diversity of biological resources and management of those resources.

Paradoxically, however, agriculture is sometimes perceived as an enemy of biodiversity. In fact, some forms of conventional agricultural growth do threaten and erode biodiversity. *(See Chapter 3.)* Yet, such conflicts are not inherent in agriculture. On the contrary, sustainable forms of agriculture can be compatible with biodiversity conservation through ways that are explored in this report. Food production depends on a variety of managed ecosystems and natural resources both in farms and in surrounding habitats such as forests, grasslands, and aquatic ecosystems. *(See Box 2.)*

Agrobiodiversity refers to the many dimensions of biodiversity that feed and nourish people and are tied to agriculture and food production at the genetic, species, and ecosystem levels. *(See Figure 2.)* The concept encompasses:

- the genetic resources that are the essential living materials of plants and animals;

- edible plants and crops, including landraces (traditional varieties), cultivars, hybrids, etc;

- livestock and edible fish or aquatic organisms;

- soil organisms that are vital to soil fertility, structure, quality, soil health, and nutrient cycling;

- naturally occurring insects, bacteria, and fungi that can attack the pests and diseases common to domesticated plants and animals;

- agroecosystem components and types (e.g., various cropping systems and landscapes, including a mix of crops, trees, livestock, soils, and topographies) that are important for productivity; and

- "wild" resources (flora and fauna) of natural habitats and landscapes that provide services (e.g., pest control or ecosystem stability) relevant for agricultural development.

Agrobiodiversity also includes the knowledge for managing biological resources—i.e., the many ways farmers *use* the biological diversity of crops, trees, soils, animals, insects, and biota for food production.[6] *(See Figure 3.)* Farmers' traditional management practices vary and are adjusted to local cultures and resources; they are therefore linked to rich cultural diversity worldwide.

Plants
There are approximately 75,000 species of edible plants globally;[7] however, over the course of human civilization, only about 7,000 plant species have been cultivated and collected for food by humans.[8] Only about 3,000 plant species (both "wild" and domesticated) are regularly exploited as food,[9] while just 103 species contribute 90 percent of the world's plant food supply.[10] Among these species, thousands of genetically distinct crop varieties have been developed through evolution and human selection and adapted to different environments and socioeconomic needs.

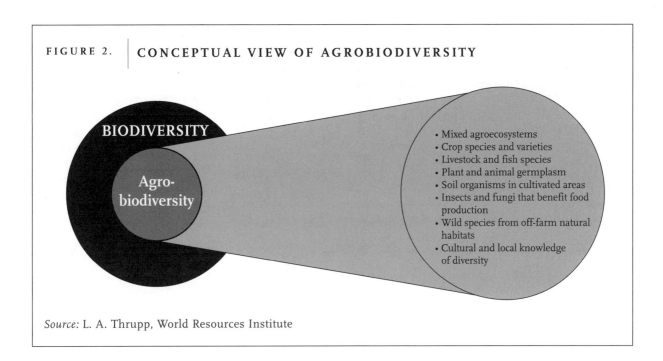

FIGURE 2.

CONCEPTUAL VIEW OF AGROBIODIVERSITY

BIODIVERSITY

Agro-
biodiversity

- Mixed agroecosystems
- Crop species and varieties
- Livestock and fish species
- Plant and animal germplasm
- Soil organisms in cultivated areas
- Insects and fungi that benefit food production
- Wild species from off-farm natural habitats
- Cultural and local knowledge of diversity

Source: L. A. Thrupp, World Resources Institute

Livestock

Livestock diversity is also important to the food supply. About 40 species of mammals and birds are recognized as domesticated species, while wild relatives of domestic livestock consist of at least 35 species.[11] Although this may seem a relatively small number, there are thousands of genetically diverse breeds among these major livestock species. Over centuries, natural evolution and selection have produced a great variety of breeds that have adapted to local conditions. Together, domestic animal species and breeds provide approximately 30 to 40 percent of the value of all food and agriculture production globally.[12] Animals also contribute work power and waste matter for fertilizing crops, and are valuable sources of income in farming systems.

Insects and Fungi

There is also a remarkable diversity among insects, fungi, and other organisms that is valuable to the productivity of agroecosystems. Arthropods, the most abundant class of animals, have an important role in biomass and agroecosystem balance. Among the arthropods, insects are one of the most significant sources of biodiversity. Although insects are often regarded as the "enemies" of food production, many insects are natural enemies to crop pests and diseases, and are invaluable for pollination, biological pest management, nutrient cycling, and biomass in the farming system[13]—all of which support productivity and greatly reduce or eliminate dependency on agrochemicals.[14]

Many kinds of fungi also contribute to the functioning of agroecosystems and crop productivity. Mycorrhizae, the various species of fungi that live in symbiosis with the roots of plants, are essential for nutrient and water uptake. They also help maintain soil fertility. Other fungi are essential for the decomposition and breakdown of nutrients and organic material.

Organisms in Soil

Soil is like a "living organism," made up of seemingly endless varieties of insects, microbes, and other microscopic organisms, including bacteria, fungi, algae, and protozoa.[15] *(See Figure 3.)* In fact, there are many more varieties of soil biota than of plants and animals above ground.[16] These organisms enhance microbial activity, increase soil fertility and aeration, accelerate decomposition, and mediate transport processes in the soil.[17]

Earthworms and other invertebrates, for example, contribute to the formation of topsoil, bringing an estimated 10 to 500 metric tons of soil per hectare to the surface per year. Some estimates put the value of such activity at $50 billion per year on agricultural land worldwide.[18] Microorganisms, small organisms such as insects and other invertebrates, help maintain nutrient cycling, soil structure, and moisture balance, as well as contributing to the natural fertility of soils.

FIGURE 3. | DIMENSIONS OF AGROBIODIVERSITY

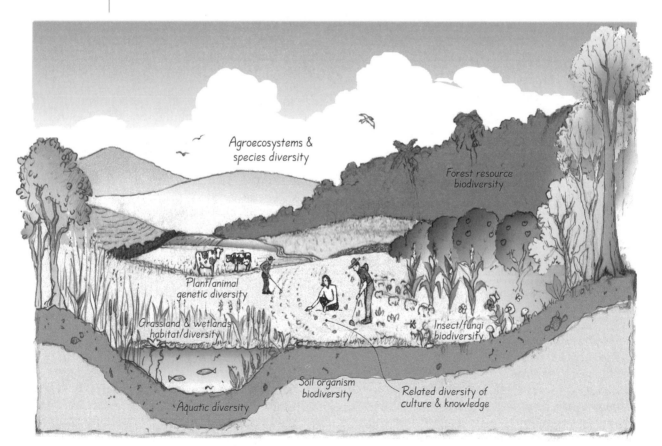

Source: L. A. Thrupp, World Resources Institute.

WRI: Cultivating Diversity: Agrobiodiversity and Food Security

Species richness in soils increases ecosystem complexity, quality, and resilience to changes[19] and is therefore valuable for maintaining productivity and enhancing soil quality—often called "soil health." In addition, diverse soil organic matter is a major source of carbon, which is valuable for regulating and mediating climate.[20] Despite these multiple benefits, the rich resources of soils are largely invisible to and unrecognized by the public.

Natural Habitats

Agrobiodiversity also includes habitats and species outside farming systems that can benefit agriculture, provide food sources, and enhance ecosystem functions.[21] Among the most important habitats tied to agriculture are forest resources and the highly diverse tree species found in forests. In many parts of the world, particularly in tropical regions, forests provide an incredible wealth of products and benefits including fodder, fiber and food sources, such as nuts, seeds, leaves, fruit, honey, roots, tubers, mushrooms, and undomesticated animals.[22] An estimated half-billion people live in or depend on forests.[23] These people's food security, nutrition, and basic health are directly linked to forests and their products. Forests also harbor genetic resources that are important for improving crops and livestock. For example, the genetic variability in red jungle fowl that live in humid forests of Asia is one of the most important sources of diversity for domestic chicken breeds.[24]

Aquatic Organisms

Aquatic organisms and ecosystems—both freshwater and coastal systems—also have rich biological wealth in terms of diverse fish species and ecosystem services. Much is still unknown about aquatic diversity in oceans, but thousands of species are known to be important for human consumption. Waters in the tropics are thought to contain the largest numbers of fish species. Brazil, for example, has more than 3,000 freshwater species, and the Indo-West Pacific Ocean has an estimated 1,500 fish species.[25]

Fish are a significant part of the world food supply, both for human consumption and as food for livestock. Globally, fish make up an estimated 17 percent of the animal protein in the human diet.[26] For millions in the South, fish are often the main source of animal protein. Fishing, fish processing, and trading provide a means of food security, jobs, and a basis of cultural traditions in coastal and inland communities.

Ecosystems

Diversity in ecosystems—both within and outside farms—provides valuable services such as water retention, nutrient cycling, and harboring of beneficial insects. These services are valuable for sustaining food production, and are also valuable for broader social interests, such as protection of watersheds and prevention of erosion. In broader landscapes or regions, agroecosystem diversity also reduces susceptibility to major climate stresses and pest or disease pressures. Diversification of cropping systems clearly has economic advantages as well, alleviating dependency on given uniform crops and varieties. Although farmers and countries have often been pushed toward monocultural and simplified farming systems, experience has shown that diversification strategies are more financially sound and bring less risk to individual producers and economies. (See Chapter 4.)

BIODIVERSITY AS A KEY INGREDIENT OF FOOD SECURITY AND SURVIVAL

Since the beginning of plant domestication, agrobiodiversity has been appreciated and nurtured by farmers as a basis of food security and survival. In the Andean region, for example, indigenous rural communities maintain and use some 3,000 varieties of potatoes from eight species. In Papua New Guinea, farmers cultivate an estimated 5,000 varieties of sweet potatoes, and include as many as 20 varieties in a single plot. In Java, farmers plant more than 600 crop species in a single home garden.[27]

The majority of staple crops consumed globally originate from a few areas—mostly in Asia, Africa, and Latin America—often called the "megadiversity"centers. *(See Figure 4.)* Although many of these crops have been cultivated by farmers around the world, crop diversity is still most concentrated in these regions. The rich diversity of crops and plant varieties here has also served as a basis for the growth of important civilizations.

Over the centuries, farmers, including many groups of indigenous peoples, have employed numerous practices to use and enhance agricultural biodiversity. They have maintained or extended the diversity of crops, livestock, trees, and "wild" flora or fauna on their farms and in the surrounding habitat, and have made use of insect and soil biodiversity. They have also undertaken their own innovations, selection, experimentation, and exchange of seeds and varieties adapting their practices to diverse environments. These traditional and dynamic forms of innovation and management are key farming practices that persist today and have produced an immense diversity of plants.[28]

Such methods have also provided many advantages for production and survival across generations. Traditional multiple cropping systems still provide as much as *20 percent* of the world food supply.[29] An estimated 60 percent of the world's agriculture is farmed by traditional or subsistence farmers, both women and men. Although these people and the areas where they live are often called "resource poor," their farming systems tend to be rich in agricultural and cultural diversity.

Although the specific components and practices of various traditional systems vary, many similarities can be found. Farmers often incorporate multiple species as well as locally adapted practices to conserve and enhance biodiversity for nutrient recycling, soil fertility, and pest management. *(See Box 3.)* Medicinal plants, valuable for health care, are often included. In many areas, farmers also inte-

---

BOX 3. | **COMMON FEATURES OF AGROBIODIVERSITY IN TRADITIONAL FARMING SYSTEMS**

Traditional farmers incorporate many aspects of agrobiodiversity including:

- a richness of plant and animal species;

- a wide diversity of niches in the local environment;

- reuse of organic residues to conserve biomass;

- enhanced ecosystem functions such as pest, weed, and disease management;

- optimal use of locally available natural resources and human resources;

- sophisticated local knowledge about plants, animals, and genetic resources; and

- cultural diversity and varying food preferences linked to agrodiversity.

These features are still used in current times in many regions of the world.

Source: Adapted from Altieri, M. 1991. "Traditional Farming in Latin America." *The Ecologist* 21(2): 93 and UNDP (United Nations Development Programme). 1995. "Agroecology: Creating the Synergism for a Sustainable Agriculture." *UNDP Guidebook Series.* New York: UNDP.

FIGURE 4. | CENTERS OF PLANT GENETIC DIVERSITY

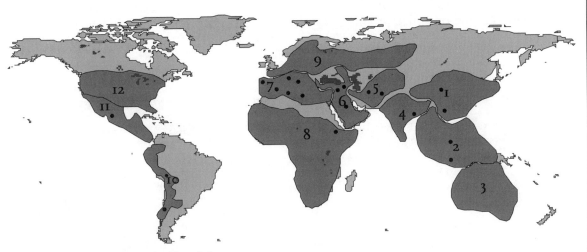

- ● Centers of origin of the principal cultivated plants
- — Gene megacenters of cultivated plants

1. CHINESE-JAPANESE REGION
   Soybean   Orange
   Rice   Tea
   Millet   Mustard
   Bamboo   Peach

2. INDOCHINESE-INDONESIAN REGION
   Banana   Rice
   Sugarcane   Yam
   Bamboo   Mango
   Coconut

3. AUSTRALIAN REGION
   Macadamia nut

4. HINDUSTANI REGION
   Rice   Mango
   Banana   Bean
   Sugarcane   Eggplant
   Cucumber   Mustard
   Chickpea   Citrus

5. CENTRAL ASIAN REGION
   Wheat   Rye
   Grape   Apple
   Apricot   Plum
   Pear   Melon
   Onion   Carrot
   Pea   Spinach
   Bean   Walnut

6. NEAR EASTERN REGION
   Wheat   Barley
   Lentil   Rye
   Grape   Almond
   Melon   Fig
   Pistachio   Pea

7. MEDITERRANEAN REGION
   Wheat   Oats
   Olive   Beetroot
   Radish   Lettuce
   Fava bean   Grape
   Cabbage   Celery

8. AFRICAN REGION
   Wheat   Sorghum
   Millet   Teff
   Yam   Oil Palm
   Coffee   Okra

9. EUROPEAN-SIBERIAN REGION
   Hops   Apple
   Pear   Cherry
   Chicory   Lettuce

10. SOUTH AMERICAN REGION
    Potato   Sweet potato
    Cassava   Tomato
    Pineapple   Cacao
    Groundnut   Lima bean
    Squash   Papaya

11. CENTRAL AMERICAN REGION
    Maize   French bean
    Potato   Squash
    Pepper/ chili

12. NORTH AMERICAN REGION
    Sunflower   Blueberry
    Jerusalem artichoke

Source: FAO (Food and Agriculture Organization of the United Nations) as cited in Shand, H. 1997. *Human Nature: Agricultural Biodiversity and Farm-Based Food Security.* Ottawa: RAFI.

grate woody species into farming systems, constituting agroforestry systems. Trees provide sources for fodder, fuel, fibers, and manures as well as food. These approaches help reduce risks and optimize productivity over the long-term.

These diverse systems and practices are not only sustainable, they also tend to be very *intensive*, in terms of maximizing use of space, plants, nutrients, animals, germplasm, and agroecological knowledge. They form a foundation for ecologically based production, and also are adapted to

---

BOX 4.   **LOCAL KNOWLEDGE AND AGROBIODIVERSITY**

Traditional Home Gardens: The Essence of Diversity

Home gardens are highly intensified farming systems that contain an extraordinary number and variety of plants. Home gardens in Mexico and Central America, for example, are known to contain at least 100 species of plants in individual plots less than one-half hectare in size.[1] Some home gardens in the Yucatan have nearly 400 plant species.[2] In Southeast Asia, similar diversity and abundance have been recorded, particularly in Indonesia.[3] Trees, shrubs, and small livestock are often included in the farming system, as well as edible, medicinal, and ornamental plants. Relay cropping, crop rotations, and living fences are common. Farmers in these systems often incorporate seeds from local natural habitats and maintain seed reserves.

The diversity within these small farms provides households with multiple sources of food and nutrition, as well as providing self-contained, vital services for soil nutrients and soil health, fodder, fibers, shade, and water retention year-round. What's more, agrochemicals are rarely used in these settings. The products of home gardens are usually for household consumption purposes, but excess crops from the gardens are often commer-

cialized as well. In one small forest-based home garden in Kalimantan, for example, a 0.2-hectare transect contained 44 species of trees, including 30 species of edible fruits or shoots, sugar palms, rubber trees, and a variety of species used for construction. The income from such a farm is substantial, because the minimum price per fruit is 15 cents. Management knowledge for these systems is passed down from generation to generation, and women are often actively involved in maintaining these plots.[4]

Traditional Farms in the Amazon: Examples From the Kayapó

The Kayapó, indigenous peoples of the Brazilian Amazon, live in an area covering about 5 million acres with a wide range of ecosystems from forests to grasslands. The Kayapó have intricate knowledge of the local soil structure, microclimate, plant and animal species, and the specific needs of each species, including their interaction.[5] They use this knowledge to create forest "islands," called *apete*, which are both agroforestry plots and hunting reserves. Kayapó management of the apete has increased both the species diversity of the managed region and the soil fertility. Management techniques include not only intercropping but the use of endemic insects for pest control. For exam-

social and economic needs. The dimensions of diversity provide benefits for productivity, food security, resilience, risk reduction, health, and income generation for local people. Being intensive rather than extensive, they also alleviate pressures on surrounding habitats in many cases.[30]

Good examples of traditional farming systems that maximize diversity are small polycultural plots, sometimes called home gardens, found in many regions including Southeast Asia, Latin America, sub-Saharan Africa, and even Europe. *(See Box 4.)* Similarly, traditional

---

BOX 4. | (CONTINUED)

ple, the *Azteca* ant, which produces a repellent chemical, is used to control leaf-cutter ants. The Kayapó also maintain "corridors" between fields, which serve as reserves for populations of various species.

Their fields are strategically planted in circular, concentric patterns that feature multicropping.[6] A circular field is cleared by selectively felling trees so that their crowns fall on the outer periphery of the circle. Then, controlled burning of these trees provides nutrients (ash) for the crops. Tubers such as sweet potatoes and yams are planted in the middle circle, followed by rice and corn, then more tubers. Taller crops like mango, cotton, banana, and papaya are planted in the outer circles. One study in Pará, Brazil, found that in a 5-year period Kayapó farming methods yielded 3 times more than colonist farming systems and 176 times the yield (in weight) of the cattle ranches.[7] With time, the trees from the surrounding forests grow onto the fields again. The total regeneration process takes anywhere from 10 to 15 years.

Notes

1. UNDP (United Nations Development Programme). 1995. "Agroecology: Creating the Synergisms for a Sustainable Agriculture." *UNDP*

Guidebook Series. New York: UNDP. p. 7 citing Alcorn, 1984.

2. Herrera. 1994. Cited in Smith, N. 1996. "The Impact of Land Use Systems on the Use and Conservation of Biodiversity." Draft paper, Washington, D.C.: The World Bank (AGRAF).

3. Padoch, C. and C. Peters. 1993. "Managed Forest Gardens in West Kalimantan, Indonesia." In C. Potter et al., eds., *Perspectives on Biodiversity: Case Studies of Genetic Resource Conservation and Development.* Washington, D.C.: AAAS Press.

4. Altieri, Miguel. 1991. "Traditional Farming in Latin America." *The Ecologist* 21(2): 93–96; Alcorn, J.B. 1991. "Indigenous Agroforestry Strategies Meeting 'Farmer Needs'."

5. Extracted by Rita Banerji from Posey, D. 1989. "Alternative to Forest Destruction: Lessons from the Mebengokre Indians." *The Ecologist* 19(6): 241–50.

6 Stevens, W.K. 1990. "Research in Virgin Amazon Uncovers Complex Farming." *New York Times,* June 3.

7 Hecht, Susan B. 1989. "Indigenous Soil Management in the Latin American Tropics: Some Implications for the Amazon Basin." In J. Browder, ed., *Fragile Lands of Latin America: Strategies for Sustainable Development.* Boulder, Colorado: Westview Press.

systems of shifting cultivation (sometimes called "swidden farming") also contain high biodiversity—i.e., a great variety of plants or livestock, both spatially and temporally. Contrary to popular perception, these systems can be relatively productive and sustainable environmentally in certain areas of the world, particularly where economic and demographic growth pressures are low.[31] Variations of shifting cultivation are still found in present day farming in many parts of the world.

Another example of sustainable use of agrobiodiversity is in traditional agroforestry systems such as shaded coffee farms in Latin America.[32] Farmers typically integrate into their coffee farms many different leguminous trees, fruit trees, and types of fuelwood and fodder. Traditional agroforestry systems commonly contain well over 100 annual and perennial plant species per field.[33] These trees provide shade, a habitat for birds and animals that benefit the farming system, biomass, nutrients, and natural insect management. For example, shade coffee plantations in Mexico support up to 180 species of birds, which are a means of pest control and seed dispersal.[34] The foliage from trees on coffee farms also helps improve soil fertility and reduce or eliminate the need for agrochemicals. The addition of diverse trees into a variety of mixed farming systems can also help restore land, as well as adding economic value.

For example, in the Machakos region of Kenya, the local people transformed degraded grassland areas and monocultural fields into diversified cropping sytems with trees and terracing, which increased production and restored land.[35]

Still another important dimension of traditional agrobiodiversity is the use of "folk varieties," also known as *landraces*, which are defined as "geographically or ecologically distinctive populations [of plants and animals] which

are highly diverse in their genetic composition."[36] Farmers have developed these unique varieties and the complex means of selecting and storing them as adaptations to local conditions. They are often valuable for resistance to pests and disease.

Equally important, many agricultural communities around the world use "wild" resources, which include nondomesticated plant and animal species found in natural habitats.[37] Wild resources are used to improve food supplies and nutrition, to supplement income, and to provide medicines and materials. A great diversity of wild resources is found in multilayered complex agroforestry systems and home gardens. Wild foods provide insurance against crop failures, and help ensure a steady supply of food year round. There are numerous documented examples of such local practices to enhance diversity in agroecosystems. A few are mentioned in *Box 4* and other recent applications are described in the final section of the report.

DIVERSITY OF KNOWLEDGE AND CULTURES

Agrobiodiversity in traditional systems is based on the *local knowledge* of people that has accumulated through experience.[38] This knowledge is by no means simple or static; it is usually complex, incorporating dynamic changes that have occurred over time. Knowledge about biodiversity is also closely tied to diverse cultural beliefs, customs, and practices. All over the world, farmers innovate, experiment, and learn to adapt and diversify agricultural practices.

Farmers in many areas are fully aware of the multiple benefits of diversity for ensuring livelihoods and reducing risks. Traditional farmers in many areas have developed their own "informal science" and methods of experimentation.[39] The Tzeltal Mayans of Mexico, for example, recognize more than 1,200 species of plants, while the P'urepechas recognize more than 900 species and the Yucatan Mayans, some 500.[40]

Both men and women possess such knowledge about agrobiodiversity. Because women are major managers or curators of farms, particularly in subsistence farming and home gardens, they have developed a sophisticated knowledge base of the characteristics of diverse crops.[41] According to several sources, rural women in many regions have unique insight, often not known to men, about diverse crop seeds and their qualities, values, and sources, especially concerning medicinal plants and landraces. They also have special knowledge about tree species and their uses in farming systems for food, fodder, and health.[42] Women in India have often transplanted many species from forests to farms and domesticated them for food crops.[43] In some cultures, the elders—both men and women—are also knowledgable about the value of diverse plants and varieties that have been used traditionally by their communities.

Local practices and special knowledge about agrobiodiversity are also often associated with rich cultural traditions and can be passed down from generation to generation. Cultural and biological diversity have evolved together, influencing each other. For example, indigenous communities in the Amazon and in Mexico celebrate cultural events, have spiritual beliefs, and perform ceremonies associated with particular practices and stages in the diverse agricultural cycle. Among the Hopi indigenous peoples in North America, different corn varieties are used in religious ceremonies.[44] For some groups, as in Zimbabwe and Ghana, wild resources around farmlands are associated with spiritual values.[45] In India, several grain crops and tree species are used in religious and festive ceremonies, inculcating the values and importance of maintaining

and using biodiversity. Such local knowledge and culture has also helped people cope with harsh conditions over time.

In some areas, certain traditional practices (such as the use of fire for clearing land) have become unsuitable because of pressures of land concentration, economic growth, demographic changes, and colonization policies. Nevertheless, traditional knowledge and practices offer lessons or principles that can enhance agrobiodiversity in a wide range of agricultural systems.

BIODIVERSITY, AGRICULTURAL SCIENCE, AND TECHNOLOGY

Agrobiodiversity has also proven valuable for large-scale commercial production, scientific and technological discoveries for crop improvement, and for increasing economic returns in farming. In particular, diverse germplasm is a crucial element for scientific and technological advances in plant and livestock breeding and improvements—and therefore is important for industrial agribusiness as well as traditional small-scale farming.

The systematic and scientific use of biodiversity has been developed and refined over the centuries. The ancient Greeks in 1600 B.C. have the earliest record of systematic plant classification.[46] Seed collecting was a central motive of explorers like Columbus. During the colonization of the "New World," collecting seeds and exploiting the genetic resources of plants was closely tied to the expansion of political control and trade. And, in the 16th and 17th centuries, with the continuation of global collection expeditions and the advent of new techniques in plant propagation,[47] explorers and scientists continued to seek control of germplasm collected in the tropics. This endeavor was closely tied to both the expansion of markets and political control.

In the 19th century, scientists discovered how to cross diverse varieties and select for certain desirable characteristics, such as pest resistance. The resulting new varieties and the insights from such experiments later became a foundation for breeding programs and scientific advancements that have enabled increases in productivity in the 20th century.[48] In addition, botanists from Russia and Europe took special interest in systematic plant collection, undertaking worldwide expeditions in the search of germplasm and the conservation of plant diversity.

Seed banks have become a major means of maintaining and storing diverse plants and germplasm from around the world.[49] As *ex situ* conservation centers, seed banks tend to be inaccessible to farming communities of the South and suffer from other deficiencies *(see Chapter 4)*, but they are useful for agricultural research institutions and for plant breeding.

Genetic diversity and access to germplasm continue to be vital for modern agriculture, plant breeding, and new methods of bioengineering and biotechnology. Diversity is valuable in the development of new medicinal products, fibers, or foods. In the United States, for example, for two major crops (soybeans and maize), exotic germplasm "adds a value of $3.2 billion to the nation's $1 billion annual soybean production, and $7 billion to its $18 billion annual maize crop."[50] The genes from wild relatives are also valuable. For instance, between 1976 and 1980, wild species contributed about $340 million per year in yield and disease resistance to the U.S. farm economy.[51] Moreover, a gene from wild tomatoes in the Peruvian Andes has increased the annual sale of the commercial tomato by $5 million to $8 million in the U.S.[52] Wild and domesticated plant species are also valuable sources for medicines, which generate considerable income. In the developing countries, more than 6,000 plants are used in traditional medicine and they are increasingly used in industrial as well as developing countries.[53]

Increasing numbers of large commercial producers recently began to recognize and profit from the benefits of agroecosystem diversity, such as using intercropping, crop rotation, and enhancing soil and insect diversity. Examples are found among major producers of grapes, vegetables, and rice in Europe, the U.S., and in parts of Asia. These producers generally are motivated to try such changes after experiencing significant losses with monocultural, uniform systems. These new approaches combine the principles of traditional farming systems with advanced principles of agroecology and biological resources. These methods have proven to be the "state of the art" for economically and environmentally sustainable agriculture. *(See Chapter 4.)*

3

AGROBIODIVERSITY LOSSES

"The most immediately vital elements of biodiversity—the parts we are using [for survival] every day and will need even more tomorrow—are severely threatened."

—Hope Shand, Rural Advancement Foundation
International, 1997

TRENDS IN GLOBAL AGRICULTURE AND FOOD PRODUCTION

Global agricultural production has increased greatly over the past 30 years. *(See Figure 5.)* At the same time, the international trade of germplasm, seeds, and diverse food products has also expanded. This growth stems from both the expansion of cultivated area (*extensification*) and the increased output per unit of land (*intensification*). Both the 20th century models of industrial agriculture and the Green Revolution promoted by the public and private sectors have strongly contributed to these trends. The goal has been to maximize yield per unit of land by using uniform high-yield varieties (HYVs)—crop varieties bred for maximizing yields—along with high inputs of agrochemicals in monocultural farming systems. These changes have contributed to food production increases, particularly in the North; however, such development patterns have also resulted in significant biophysical and socioeconomic problems. In particular, agrobiodiversity has been eroded or lost, and other natural resources have been degraded—which ultimately undermines productivity and decreases food supplies.

Meanwhile, food insecurity is widespread and some 800 million people suffer from hunger worldwide.[54] In sub-Saharan Africa alone, the number of chronically undernourished people has more than doubled since 1970, from 96 million to over 200 million.[55] *(See Figure 6.)* In recent decades, food security and growth rates of agricultural productivity have actually declined in many developing countries.[56] *(See Figure 7.)* In addition, resources and food are distributed highly inequitably within nations, in regions, and across the world. The poor, particularly women and children, generally suffer the main burden of these problems.

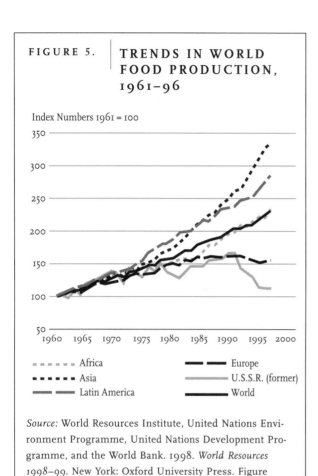

FIGURE 5. | **TRENDS IN WORLD FOOD PRODUCTION, 1961–96**

Index Numbers 1961 = 100

- - - - - Africa
■ ■ ■ ■ Asia
— — Latin America
— — Europe
U.S.S.R. (former)
— World

Source: World Resources Institute, United Nations Environment Programme, United Nations Development Programme, and the World Bank. 1998. *World Resources 1998–99.* New York: Oxford University Press. Figure FW.3, p. 154.

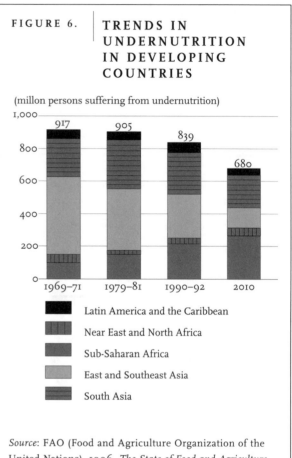

FIGURE 6. | **TRENDS IN UNDERNUTRITION IN DEVELOPING COUNTRIES**

(millon persons suffering from undernutrition)

■ Latin America and the Caribbean
▥ Near East and North Africa
▤ Sub-Saharan Africa
▥ East and Southeast Asia
▤ South Asia

Source: FAO (Food and Agriculture Organization of the United Nations). 1996. *The State of Food and Agriculture, 1996.* Rome: FAO, Figure 13, p. 271.

These trends pose tremendous challenges to meet growing food needs while conserving resources. They suggest the need for changes and improvements in sustainable food production, equitable distribution systems, and appropriate use of resources.[57] An important dimension of the challenge is the need to overcome the serious threat from the erosion of agrobiodiversity.

LOSS OF DIVERSITY IN AGRICULTURE

Agrobiodiversity has been deeply eroded throughout the world. At the same time, productivity is precarious at best and food security is threatened—all of which have serious implications for humanity. The losses and their impacts are manifested in many ways and at different

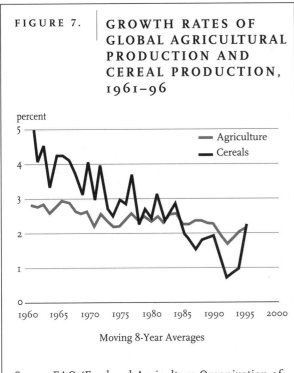

FIGURE 7. | **GROWTH RATES OF GLOBAL AGRICULTURAL PRODUCTION AND CEREAL PRODUCTION, 1961–96**

percent

— Agriculture
— Cereals

Moving 8-Year Averages

Source: FAO (Food and Agriculture Organization of the United Nations). 1997. *FAOSTAT Statistical Database.* Rome: FAO.

"wild" animal species, losses affecting agriculture have severe implications for global food security.[58] Many domesticated local varieties and their wild relatives are being lost as the result of homogenization of farming, the spread of uniform varieties, and replacement of polycultures by monocultures. Fewer than 100 species currently provide most of the world's food supply, even though thousands have been grown since the dawn of agriculture.[59] Currently, just three crops—rice, wheat, and maize—account for about 60 percent of the calories and 56 percent of the protein that people derive from plants.[60]

Since the arrival of Columbus in the Americas, an estimated 75 percent of the native food crops have been lost in the hemisphere. In Mexico, for example, only 20 percent of the maize varieties reported in 1930 are currently used there.[61] In China, between 1949 and 1970, the number of wheat varieties cultivated by farmers dropped from about 10,000 to only 1,000. Between 1910 and 1920 in Taiwan, the number of landraces of rice dropped from 1,200 to approximately 400 as a result of a campaign against rice diversity; since then, the numbers have fallen still more dramatically.[62]

levels, both within farming systems and off farms in natural habitats such as forest lands. The poor tend to bear the greatest burden and costs from these losses. These different types of threats emanate from common root causes— linked to prevailing paradigms, conflicting policies, and inappropriate food production practices. *(See "Underlying Causes," below.)*

Genetic Diversity
Genetic erosion among plant and animal species is occurring at a rapid rate and is one of the major concerns of agrobiodiversity loss. (Genetic erosion means loss of genetic diversity between and among populations of the same species.) Although not as well-publicized as the loss of

Although plant breeders draw upon a very wide diversity of plants to infuse many genetic traits into new varieties, the resulting seeds/varieties sold and recommended to farmers tend to be uniform.[63] Thousands of traditional crop varieties have been eliminated from use. The use of wild species is also declining worldwide, and some of these are close relatives of cultivated varieties with a potentially high value as medicines or food.

In Bangladesh, for example, the promotion of HYV rice monoculture has led to major losses of diversity, including nearly 7,000 traditional rice varieties and many fish species. The per-acre production of HYV rice in 1986 dropped by 10 percent from 1972, despite a 300-percent increase in agrochemical use per hectare.[64] In the Philippines, the introduction of HYVs has displaced more than 300 traditional rice varieties that had been the principal source of food for generations. In India, by 1968, the so-called "miracle" HYV seed had replaced half of the native varieties; however, these seeds were not "high-yielding" unless they incorporated high fertilizer inputs and irrigation, which were often unaffordable for poor farmers, so expected production increases were not realized in many areas.[65]

African nations and communities have also suffered from significant declines in agricultural diversity. Traditional foods and varieties, such as teff in Ethiopia, have been replaced by exotic varieties of wheat and maize developed for different ecosystems. In Senegal, for example, a traditional cereal called fonio (*Panicum Laetum*)—which is highly nutritious as well as robust in lateritic soils—has been threatened by extinction because of the replacement by modern crop varieties.[66] In the Sahel, reports also confirm that traditional systems of polycultures are being replaced with monocultures. These changes cause further food insecurity and also contribute to economic decline.[67] *(See Table 1 for further examples.)*

This kind of homogenization happens in high-value export crops as well. Nearly all the coffee trees in South America are descended from a single tree from a botanical garden in Holland, *Coffea arabica*, which was first obtained from the forests of southwest Ethiopia—forests that have virtually disappeared.[68] Unfortunately, traditional coffee farms with a high diversity of shade trees and fauna that are advantageous for productivity have been replaced by new treeless uniform coffee systems. Uniform varieties are also common in export crops such as bananas, cacao, and cotton, replacing traditional diverse varieties. Such changes may have generated short-term productivity increases, but they also create significant risks and losses over time.

TABLE 1. | **DECLINES IN AGRICULTURAL DIVERSITY, SELECTED CROPS**

Crop	Country	Number of Varieties	Source
Rice	Sri Lanka	75% of varieties are descended from a common stock. Down from 2,000 varieties in 1959 to less than 100 today.	Rhoades, 1991
Rice	Bangladesh	62% of varieties are descended from a common stock.	Hargrove et al., 1988
Rice	Indonesia	74% of varieties are descended from a common stock.	Hargrove et al., 1988
Wheat	United States	50% of crop in 9 varieties.	NAS, 1972
Potato	United States	75% of crop in 4 varieties.	NAS, 1972
Soybeans	United States	50% of crop in 6 varieties.	NAS, 1972

Source: World Conservation Monitoring Centre. 1992. *Global Biodiversity: Status of the Earth's Living Resources.* Chapman & Hall: London.

Similar losses in crop diversity are occurring in the North. A recent survey found that the majority of fruit and vegetable varieties listed in 1903 by the U.S. Department of Agriculture are now extinct. Of more than 7,000 apple varieties used in the United States between 1804 and 1904, 86 percent are no longer in cultivation; of 2,683 pear varieties, 88 percent are no longer available.[69] *(See Table 2.)* Evidence from Europe shows similar trends; thousands of varieties of flax and wheat vanished following the takeover by HYVs.[70] Varieties of oats and rye are also declining in Europe.[71] In Spain and Portugal, for example, various legumes that had been an important part of the local diet are being replaced by homogeneous crops. In Holland, four crops are grown on 80 percent of Dutch farmlands.[72]

Livestock species are also suffering genetic erosion. The FAO estimates that 30 percent of livestock worldwide are classified as endangered and critical. It also reports that at least one breed of traditional livestock is lost every week somewhere in the world.[73] Many traditional breeds have disappeared as farmers focus on producing new breeds of cattle, pigs, sheep, and chickens.[74] Of the 3,831 breeds of cattle, water buffalo, goats, pigs, sheep, horses, and donkeys that have existed this century, 16 percent have become extinct, and a further 15 percent are rare.[75] Some 474 of extant (livestock) breeds can be regarded as rare. A further 617 have become extinct since 1892.[76] More than 80 breeds of cattle are found in Africa, and some are being replaced by exotic breeds.[77] In Europe, half of the breeds of domestic animals that existed at

---

TABLE 2. | VARIETIES IN THE U.S. NATIONAL SEED STORAGE LABORATORY COLLECTION, 1903 AND 1983

Vegetable	Latin Name	Number of Varieties, 1903	Number of Varieties, 1983	Loss (percent)
asparagus	*Asparagus officinalis*	46	1	97.8
beans	*Phaseolus vulgaris*	578	32	94.5
beets	*Beta vulgaris*	288	17	94.1
carrot	*Daucus carota*	287	21	92.7
leek	*Allium ampeloprasum*	39	5	87.2
lettuce	*Lactuca sativa*	497	36	92.8
onion	*Allium cepa*	357	21	94.1
parsnip	*Pastinaca sativa*	75	5	93.3
pea	*Pisum sativum*	408	25	93.9
radish	*Raphanus sativus*	463	27	94.2
spinach	*Spinacia oleracea*	109	7	93.6
squash	*Cucurbita spp.*	341	40	88.3
turnip	*Brassica rapa*	237	24	89.9

Source: Fowler, Cary and Pat Mooney. 1990. *The Threatened Gene: Food, Politics, and the Loss of Genetic Diversity.* Cambridge, UK: The Lutworth Press.

the turn of the century have become extinct; and 43 percent of the reminaing breeds are defined as "endangered."[78] These losses not only undermine the food security of people who depend on livestock as a source of food and nutrition, but they also weaken breeding programs that could improve livestock hardiness.

Freshwater and marine fish species are also imperiled as freshwater ecosystems and marine resources are increasingly degraded and depleted both by intensive commercial fishing operations and pollution from agriculture and industrial sources. At least one fifth of all freshwater fish are extinct or seriously endangered.[79] An estimated 70 percent of the world's marine fish stocks are either depleted, overfished, recovering from overfishing, or fully exploited.[80] Fish catches have been declining or stagnating globally. In four Atlantic fisheries and one Pacific fishery, for example, total output of fish has dropped more than 30 percent.[81] This is a serious threat for people who depend on local fisheries and harms thousands more who have lost their jobs from declining catches.

Agroecosystem Services

Related to the loss of germplasm, seed stock, and livestock varieties are broader losses of diversity in farming systems and landscapes. In many countries, diverse farming systems and practices, such as polycultures, intercropping, and agroforestry have been displaced or have even disappeared, largely due to the spread of monocultural systems and industrial agricultural methods. Farmers around the world have been encouraged and sometimes obligated by extension programs, credit and agricultural policies, and food buyers and marketers to replace traditional diverse practices with new uniform varieties, standardized livestock breeds, and monocultural models. This erosion of ecosystem diversity degrades valuable ecosystem services, such as water management, nutrient cycling,

soil fertility, and organic matter, thus hindering productivity and sustainability.

In the Kolli Hills of South India, for example, people have been persuaded by agroindustries to plant tapioca in monocultures instead of fruit trees and mixed crops, the major components of their indigenous agroforestry systems. Although the tapioca provides immediate income, it has had negative ecological and economic effects including reduced availability of water, soil depletion, and increased expenditures and debts.[82] Throughout Latin America during the 1980s, as a result of policy mandates, coffee producers replaced traditional diverse agroforestry cofffee systems with "modern" varieties that require elimination of trees and other diverse vegetation. (However, this change led to significant risks and losses in coffee plantations and was later recognized as a costly mistake.)

In Costa Rica's Guanacaste region, when an expansive project of irrigated rice was developed in the mid-1980s, all of the local farmers who traditionally planted diverse polycultural farms (with corn, beans, vegetables, root crops, and fruit trees) had to either convert to monocultural high-yield rice or vacate the area.[83] While the rice plantations brought new income opportunities for large producers, the transformation displaced people who had made a living from sustainable integrated systems.

Increased Vulnerability to Pests and Diseases

One of the most serious problems associated with the loss of diversity in crops and livestock is the increased vulnerability to pests and diseases. A plant pest or disease can be devastating if it infests a uniform crop, especially in large, homogeneous plantations. Producers have suffered serious economic losses from relying on monocultural varieties. Well-known examples include the great potato famine in Ireland in the 19th century that caused massive starvation in the United

Kingdom, a wine grape blight that wiped out valuable vines in both France and the United States, a virulent disease that damaged expansive banana plantations in Central America in recent decades, and devastating mold that infested hybrid maize in Zambia. *(See Box 5 and Table 3.)* The threat from such penetrating pest devastation will continue into the future as long as monocultural systems prevail. Uniform commercial potato production in western industrial nations is currently jeopardized by late potato blight, the same fungus that caused the potato famine in Ireland. Late blight is threatening the $160 billion potato industry in the United States, and is causing yield losses of up to 30 percent in Third World potato crops, especially in those vulnerable areas where potato diversity has been lost.[84]

In each case, losses probably could have been avoided if diversity of crop varieties and biodiverse food production systems had been used. Scientists and government agencies including FAO and the U.S. National Academy of Sciences acknowledged the vulnerability arising from uniform stock as far back as the 1970s.

Disruption of Insect Diversity

The diversity of insects and fungii is likewise being eroded with a resulting loss of important ecosystem services such as pollination, natural control of crop pests and diseases, and nutrient recycling. This leads to increasing costs, crop losses, and declining productivity on farms. The destruction of insect diversity also opens the door for a resurgence of pests or outbreaks of new pests. Farmers then apply heavier doses of pesticides or change products. This does not solve the problem, however, and creates greater pesticide resistance in hundreds of insect species. This cycle, known as the "pesticide treadmill," has caused enormous costs and devastating crop losses in many circumstances, as in the cases of cotton and banana production in Latin America and rice in Southeast Asia.[85]

Agrochemicals generally kill natural enemies and beneficial insects on farms and in surrounding habitat as well as the target pests. Especially when overused, pesticides destroy a wide array of susceptible species in the ecosystem while also changing its normal structure and function.[86] These impacts hinder production and consequently contribute to decreases in food supplies.

Soil Biodiversity Loss

One estimate suggests that soil is being lost worldwide at a rate 13 to 80 times faster than it is being formed.[87] The erosion of soil biodiversity undermines soil fertility, harms soil structure, and leads to productivity losses.[88] The weaknesses in soil quality and structure can also aggravate crop susceptibility to diseases and pests. Such effects ultimately contribute to declines in the local food supply.

Recent losses in soil biodiversity and soil health can be traced primarily to unsustainable soil management practices in conventional agriculture, including the heavy use of agrochemicals (particularly pesticides, soil fumigants, and chemical fertilizers), which can destroy or disrupt soil organisms and soil quality; homogenization of crops and varieties (i.e., planting a uniform crop over time and in a given area), which depletes the soil of natural nutrients; the erosion of soils and lack of soil conservation methods; and the decline in use of natural manures and crop residues, intercropping, cover crops, and other practices that enrich soil health.

LOSS OF HABITAT DIVERSITY

Modifying natural systems to fulfill food needs is a necessary process.[89] When this can be accomplished in suitable areas and by appropriate means, agricultural development does not necessarily produce negative effects on the natural systems. However, some conventional agricultural practices lead to biodiversity losses in

BOX 5.

CRISES FROM PESTS IN VULNERABLE UNIFORM MONOCULTURES

During the Irish potato famine of 1845, the entire nation's potato production collapsed as a direct result of the country's reliance on monocultural production of a narrow stock of potato varieties that descended from the Americas in the late 15th century. This uniform stock was highly susceptible to the blight (*Phytophthora infestans*), and the resulting loss of the crop led to the death of 1 million people and the migration of 1 million more.[1]

During the 19th century in France, wine-grape production was wiped out by a virulent disease (*Phylloxera vitifoliae*) that attacked the roots of uniform varieties of grapes being grown. Four million acres of vineyards were decimated, resulting in losses of untold millions of dollars. Wine-grape growers had no choice but to pull out the plants by their roots, and later plant new varieties; but those that planted monocultures experienced a similar infestation again. Moreover, the pattern continued in the United States, in the Napa and Sonoma counties of California, home of the most valuable vineyards in the nation. Here, wine production is again under threat because 70 percent of the crop is grafted onto a homogenous rootstock that is

being infested by a new variant of the disease *Phylloxera*.[2] Wine-grape growers are again forced to rip out vines, which is costing millions of dollars. Diversification of varieties (along with nonchemical soil and crop management methods) has proven to be a critical path for reducing this susceptibility.

In Central America, the extensive banana industry has been seriously jeopardized because of a reliance on uniform varieties grown in massive monocultural plantations. Panama disease, a virulent fungus (*Fusarium oxysporum*), attacked and destroyed thousands of hectares of monocultural (Gros Michel) bananas in the Atlantic region in Costa Rica during the 1930s; this led the nearly bankrupt industry to abandon the land, move across the country, and plant a new variety (i.e., Valery or Cavendish). But when growers again planted vulnerable monocultures, the plantations again suffered from serious diseases—first Yellow Sigatoka during the 1950s, and then Black Sigatoka from the 1970s through the 1990s—costing the industry millions of dollars in losses, as well as exorbitant costs for control efforts. Banana companies continue to use a chemical-based

natural habitats, resulting in high social costs. In particular, some practices have contributed to natural habitat loss in forest areas, grasslands, and wetlands. This happens mainly through *extensification* of farming systems into the "frontier" zones as natural vegetation is cleared for production. A significant proportion of global deforestation has been attributed to agricultural expansion, often spurred by settlement policies and economic pressures.

In the conversion of "wild" habitats to agriculture, natural plant species are replaced by introduced species. This change does not necessarily result in major biodiversity losses in natural habitats because some agricultural approaches, such as polyculture and agroforestry systems or other forms of mixed cropping, conserve some natural species and functions and even enhance the diversity of species while serving production purposes.

BOX 5. | (CONTINUED)

approach to struggle against these diseases, but as long as the uniform expansive plantation systems are maintained, they are unlikely to escape the problem.[3]

In 1972, following the widespread adoption of uniform wheat HYVs, Brazil lost half its national wheat crop when it was attacked by disease.[4] Highly uniform HYVs of maize in Zambia experienced devastation when a mold infested the crop in 1974, and destroyed 20 percent of the hybrid plants. The impact on traditional maize varieties was negligible.[5] In the United States, in the early 1970s, uniform high-yielding corn hybrids comprised about 70 percent of all the corn varieties at that time—making them very susceptible to disease; and indeed, a corn leaf blight resulted in the loss of 15 percent of the entire crop in that decade. Likewise, coffee rust has destroyed uniform coffee farms repeatedly in Central America, Sri Lanka, India, Malaysia, the Philippines, and half a dozen African countries—largely due to their dependence on single varieties.[6]

Despite the lessons from history, the problems created by dependency on a monoculture have occurred again and again.

Notes

1. Fowler, Cary and Pat Mooney. 1990. *Shattering: Food, Politics, and the Loss of Genetic Diversity.* Tucson: University of Arizona Press. pp. 43–45.

2. National Research Council. 1993. *Sustainable Agriculture and the Environment in the Humid Tropics.* Washington, D.C.: National Academy Press.

3. See L.A. Thrupp. 1988. "The Political Ecology of Pesticide Use in Central America: Dilemmas in Banana Production of Costa Rica." Ph.D. dissertation, University of Sussex, for overview of the history of diseases/pests.

4. Cited in Mooney, Pat. 1979. *Seeds of the Earth: A Private or Public Resource?* Ann Arbor, Michigan: Canadian Council for International Cooperation. p. 43.

5. John, P. 1974. "The Green Revolution Turns Sour." *The Ecologist.,* pp. 304–305, cited in Mooney, Pat. 1979. *Seeds of the Earth: A Private or Public Resource?* Ann Arbor, Michigan: Canadian Council for International Cooperation. p. 13.

6. Mooney, Pat. 1979. *Seeds of the Earth: A Private or Public Resource?* Ann Arbor, Michigan: Canadian Council for International Cooperation. p. 13.

On the other hand, the conversion of frontier areas to monocultural farming systems and the use of conventional practices often reduces the biodiversity of flora and fauna in habitats and landscapes. For example, both the widespread conversion of forests to monocultural pastures and the introduction of uniform livestock breeds in the Americas and parts of Africa has prompted a significant decline in biodiversity.[90] (These trends are also tied to other adverse impacts on natural resources, such as soil erosion and water depletion.)

Increasing reliance on chemically intensive practices has also contributed to off-farm losses in natural habitats. Pesticide residues inevitably drift into surrounding air, water, and soil where insects and other flora and fauna are killed or harmed. Heavy use of chemical fertilizers usually results in runoff that pollutes habitats, soil,

TABLE 3. | CROP FAILURES DUE TO GENETIC UNIFORMITY

Date	Location	Crop	Effects	Source
1846	Ireland	potato	potato famine	Hoyt, 1988
1800s	Sri Lanka	coffee	farms destroyed	Rhoades, 1991
1943	India	rice	Great Famine	Hoyt, 1988
1960s	United States	wheat	rust epidemic	Oldfield, 1984
1970	United States	maize	$1 billion loss	NAS, 1972
1970	Philippines and Indonesia	rice	tungo virus epidemic	Hoyt, 1988
1974	Indonesia	rice	3 million tons destroyed	Hoyt, 1988
1984	United States (Florida)	citrus	18 million trees destroyed	Rhoades, 1991

Source: World Conservation Monitoring Centre et al. 1992. *Global Biodiversity: Status of the Earth's Living Resources.* Brian Groombridge, ed. Chapman & Hall, London.

and water supplies (including streams, lakes, reservoirs and groundwater) and damages human health and ecosystems while also disturbing the diversity of flora and fauna. The adverse impacts of heavy chemical use can spread widely throughout natural ecosysems and communities near farming areas.[91]

The main impacts of habitat loss on species diversity include:

- eliminating habitat use by species followed by loss of wild species;

- removing vegetation, which can affect breeding areas, reduce shelter and food sources, and result in changes in species composition;

- fragmented habitat (e.g., patches of intact or degraded habitat), which can harm ecosystems by changing nutrient or microclimatic regimes and species composition. In Brazil, for example, 39 percent of the habitat affected by agricultural expansion was completely converted to crops or pasture;[92] and

- reducing the rate of forest regeneration, partly because parent material of seeds is destroyed.[93, 94]

Not only do these changes come with high social costs (loss of nutrients, water and insect mangement, and loss of valuable wild habitat products such as medicines, natural fibers, fodder, and food), this loss of biodiversity also has moral implications for all life on the planet.[95, 96] There is harm not only to farms or communities in a locality, but also to wider regional and global ecosystems. Future as well as present generations are affected, since valuable stores of genetic resources in natural habitats may be irretrievably lost.[97]

The concerns about biodiversity loss from agricultural extensification often focus on tropical moist forests, largely because of the unusually high concentration of species in such areas.

Although tropical moist forests cover about 7 percent of the Earth's land surface, they are believed to harbor more than half of the world's plant and animal species—totalling somewhere between 2 million and 20 million.[98] Local people also depend on such biological resources. The biodiversity in these forest areas has intrinsic ecological benefits which are difficult to measure economically.[99] Yet, the number of species is declining rapidly as forests are cut down. Such losses of biodiversity are not confined to tropical forests, but are particularly relevant in relation to agricultural changes. The impacts are not exclusively off-site externalities, however. They can also directly undermine agricultural production and profits as they disrupt the ecosystem services and eliminate natural products of habitats.

Agrobiodiversity losses harm both public and individual resources and therefore pose major challenges conceptually and practically. The distinction between public and private resources becomes blurred, and the causes and effects of "external" and "internal" costs cannot be clearly distinguished.

HUMAN IMPACTS

Effects on Food Security and Nutrition
Both food security and insecurity are closely linked to environmental (biophysical) and socioeconomic conditions; and biodiversity is one of the most important physical factors directly affecting food supplies. Productivity losses stemming from declining diversity can directly reduce household food security by decreasing yields, increasing risk, and reducing food availability. In addition, food security can be jeopardized if the lack of diversity in a given system makes a farmer or region more susceptible to climate fluctuations, market price swings, or pest attacks. When communities lose diverse, locally adapted varieties, they often lose their

immediate source of food supply. In addition, the narrowing and disappearance of germplasm impedes the discovery of new crop and animal varieties that provide an important potential source of future food supplies. Poor people tend to be particularly vulnerable to these losses, because they are more dependent on diversity as a basis for livelihood and for risk reduction. Likewise, the poor generally bear the higher costs from agrobiodiversity erosion.

While many other physical, technological and socioeconomic factors contribute to food insecurity (not the least of which is the inequitable distribution of resources), the decline of biodiversity and the increasing control of germplasm by large companies are certainly major factors that profoundly aggravate hunger and food deficiencies.

One direct human impact of the decline of diversity in crops and varieties is the decline in the variety of foods consumed by people, which has adverse effects on nutrition. Among poor farmers, for example, conversion from polycultural to monocultural systems usually means that the household must rely on purchased foods that are often unaffordable or inaccessible. More generally, as people consume more uniform and standardized foods, they do not get the variety of vitamins and minerals that are important to nutrition and health. The decline and extinction of landraces and "wild" foods also means a decline in valuable nutritional sources.[100]

Much of agricultural research and development has focused on commercial or cereal production, almost to the exclusion of locally grown vegetables and legumes. This is a serious problem in countries where people depend on legumes for protein. In India, for example, chickpea acreage dropped to half that of wheat during the 1960s and 1970s when, historically, the two crops had equal acreage. Per capita legume production dropped by 38 percent between 1961 and 1972. Thus, high-protein legumes are replaced by low-protein grains. Similar changes are occurring in East Africa, where there is pressure for producers to replace indigenous traditional grains, such as teff and fonio, with exotic maize and wheat varieties that have inferior nutritional value.

Although growing international trade in agricultural products has enabled Northern consumers to eat a wider diversity of foods in some areas, at the same time, these changes in global food systems obligate farmers worldwide to produce uniform varieties that are demanded in the markets. For example, farmers in the South who grow export crops for Northern consumers are usually obligated to grow monocultures of certain varieties of fruits and vegetables that are exotic to their local conditions and foreign to their tastes. They also often replace their own diverse subsistence crops. As a consequence, nutrition among a large majority can be harmed.[101]

Displacement of Local Knowledge, Culture, and Inequitable Control of Resources

A related important concern is the loss of local knowledge about agrobiodiversity, which results from the widespread use of uniform industrial agricultural technologies. This process can create both economic and cultural losses for local people and for society more generally. Likewise, the pervasive homogenization process displaces and disrupts valuable cultural diversity. This situation is illustrated clearly by the decline of knowledge about agroforestry (in which trees have been removed from systems, as in shade coffee plantations); replacement of traditional nonchemical pest management methods by chemicals; and the loss of knowledge of medicinal plants and landraces, which have been overtaken by uniform HYVs and other products.

Another important dimension of the problem is the inequitable control and distribution of benefits of diverse genetic resources, between the North and South.[102] Although the large centers of diversity and the majority of resources are located in the developing world, the industrialized countries and transnational companies have increasing control over these valuable resources, and likewise gain more benefits from them. Meanwhile people in the South are often losing the benefit of these resources. In addition, seed banks and other sources of new germplasm are often inaccessible to poor people in the South.

UNDERLYING CAUSES

"We must identify the driving forces and eliminate the incentives leading to the destruction of agricultural biodiversity."

—Indian delegate to the Conference of Parties, Convention on Biological Diversity, May 1998

Agrobiodiversity losses have complex causes, some of which have been been briefly mentioned above. However, to find effective solutions, the underlying causes need to be identified and explored more explicitly. Losses are frequently attributed inappropriately to farmers' behaviors, particular technologies, or a lack of environmental awareness, when, in fact, the underlying forces are largely political, economic, and social.

The root causes of declines in agricultural diversity can be traced to the following:

• the dominance of industrial agricultural paradigms, policies, and institutions (including credit, research, extension, and development institutions) that support homogeneous systems and unsustainable practices,

- global inequities in the control and distribution of resources,

- pressures from businesses and market growth that promote uniform monocultures and the related packages of agrochemical technologies,

- the undervaluation of biodiversity and disrespect of local knowledge tied to such diversity, and

- policy-induced demographic pressures. *(See Table 4.)*

Paradigms and Policies

One of the most influential root causes of agrobiodiversity losses, both past and present, is the strong promotion and spread of the industrial agriculture paradigm and the closely related Green Revolution, which extended the Northern model into developing countries. This model focuses on maximizing agricultural yields and productivity through the use of monoculture systems and uniform technologies, including high-yield variety seeds (HYVs), agrochemicals, irrigation, and mechanized equipment. In recent years, agricultural biotechnologies, such as herbicide-resistant crops, have been included in this model. An enormous institutional structure, including many international donors and development agencies, international and national research institutions (including the system of the Consultative Group on International Agriculture Research (CGIAR)), and national governments, have supported and subsidized influential programs and policies to promote and spread these modern uniform varieties and the related packge of technologies.

> *"Loss of biodiversity is frequently characterized as an environmental problem, but the underlying causes are social, political, and economic."*
>
> —Hope Shand, Rural Advancement Foundation International, 1997

Major universities around the world, particularly the large agricultural colleges in the United States, have fuelled the industrial and Green Revolution model. As a result, HYVs are now used on 52 percent of the agricultural land planted in wheat, 54 percent of land planted in rice, and 51 percent of land planted in maize.[103] This has contributed to production increases in many areas of Asia and Latin America.

However, the global dominance or imposition of this model of agricultural technology growth (and the influential policies, institutions, businesses, and programs that uphold it) has provoked unexpected social costs and production problems, partly because it has directly reduced agricultural biodiversity and has led to the widespread use of technologies that have proved to be unsustainable and degrading over time. Moreover, the new varieties and related agrochemicals frequently displace diverse local indigenous varieties and practices and are poorly adapted to local agroecological conditions. The model is also highly specialized and reductionist, oriented towards producing a very narrow range of strains and varieties for highly controlled environments and intensive production. And, while adoption of the model can increase the productivity of a given specialized crop in the short term, it often increases vulnerability to pests and to climate and economic changes, raising farmers' risks and hindering

TABLE 4. | CAUSES OF BIODIVERSITY LOSSES LINKED TO AGRICULTURE

Problems	Proximate Causes	Underlying Causes (for all problems)
Erosion of genetic resources (livestock and crops/plants): — threatens food security, — increases risks, — prevents future discoveries.	Dominance of uniform HYVs and monocultures, biases in breeding methods, weak conservation efforts.	• Industrial/Green Revolution paradigm that stresses uniform monocultures. • Inequitable distribution of land and resources.
Erosion of insect diversity: — increases susceptibility to pest and diseases — ruins pollination and biocontrol.	Heavy use of pesticides, use of monocultures/uniform species, degrading habitats harboring insects.	• Policies that support uniform HYVs and chemicals (e.g., subsidies, credit policies, and market standards). • Pressures and influences of seed/agro-chemical companies and extension systems.
Erosion of soil diversity — leads to fertility loss — reduces productivity.	Heavy use of agrochemicals, degrading tillage practices, use of monocultures.	• Trade liberalization and market expansion policies that neglect social and ecological factors.
Loss of habitat diversity (including wild crop relatives).	Extensification in marginal lands, drift/contamination from chemicals.	• Lack of awareness of agroecology and R&D and education institutions.
Loss of indigenous methods and knowledge of biodiversity.	Spread of uniform "modern" varieties and technologies.	• Disrespect for local knowledge. • Demographic pressures.

productivity. In spite of the costs, the paradigms of industrial agriculture and the Green Revolution continue to dominate development paths.

In addition, numerous policies—ranging from general agricultural development policies to pricing and credit packages—directly and indirectly influence biodiversity in agriculture. Among the most influential are incentive policies (e.g., subsidies for agricultural inputs,

extension programs, credit policies, and marketing standards) that support the adoption of Green Revolution technologies. These policies exist at international, national, and local levels, and affect farmers' incentives and decisions at the local level. In many countries, for example, HYV seeds and other related technologies are priced low and subsidized, which is an obvious inducement to farmers to adopt them. In addition, agricultural credit policies often require farmers to purchase and use HYVs and agrochemicals in order to qualify for loans.[104] Extension programs in many countries have mandated the adoption of uniform varieties and the elimination of diversity, as in the case of coffee production throughout Latin America, and in recent "Green Revolution" initiatives in East Africa.[105] Such policies have led to the widespread prescription and adoption of uniform seeds and monocultures and the intensive use of agrochemicals.

Land tenure policies and colonization programs also lead people to farm in frontier areas that can degrade biodiversity. Many governments have established policy incentives for people to clear forested land and establish farms in order to gain tenure.[106] Land settlement policies in a variety of countries, including Brazil, Costa Rica, and Indonesia, promoted the expansion of agriculture into frontier zones.[107] While these policies have contributed to increases in food production and have helped producers (especially medium- and large-scale farmers) gain benefits in the short term, they often have led to land development that is unsustainable.[108] The "public good" character of resources also leads to underinvestment and overexploitation of biological resources.[109]

Corporate Influence
Private agribusinesses, particularly transnational corporations, that market agricultural inputs, have a strong influence on the research, development, and distribution of seeds and other technologies that directly affect agrobiodiversity. Before the 1940s, public institutions were mainly responsible for research and development on plant genetic resources; but during the 1940s and 1950s, this moved largely into the hands of the private sector.[110] The seed industry was initially formed by small rural enterprises, mainly family businesses that produced seeds through home operations, but seed companies rapidly grew into a multibillion-dollar global

industry. In the 1960s, companies dealing with petroleum-based products entered into seed production and gained commercially by consolidating both the seed and agricultural input businesses.[111] Over the next 20 to 30 years, the influence of the large chemical companies increased with further consolidation in the seed and breeding business, as well as in agrochemicals.[112] Over the past two decades, the large companies dedicated to seed and agrochemical sales have also become increasingly concentrated among a few transnationals, while also taking over biotechnology development. These companies depend on access to a wide variety of germplasm for breeding, engineering, and innovations that they control through the use of patents.[113] They have major influence over the direction of agricultural technology development, by controlling and widely marketing uniform HYVs, chemicals, and biotechnologies that are generally oriented toward homogenized, narrow production systems.

In recent years, some corporations have stressed biotechnology inventions for pesticide-resistant crops and seeds that terminate germination after one growing season. Such biotechnologies aggravate these threats to biodiverse and sustainable cropping systems. Therefore, they have generated major controversy and opposition among consumers, scientists, policy agencies, and farmers in many countries.

Market Pressures and Undervaluation
Even though biodiversity has many benefits, it is undervalued or even ignored in conventional economic assessments.[114] This occurs in part because the benefits of biodiversity (particularly ecosystem services) are difficult to value in economic terms. In addition, there is a lack of complete understanding of how to measure these benefits and advantages for both present and future purposes. The value of practices such as using mulch or cover crops to enhance soil

diversity or conserving indigenous seeds can not be easily measured in terms of production and sustainability.

Conversely, the costs from losses of biodiversity, such as the build up of pests, destruction of beneficial insects, or of ecosystem services, are also difficult to value. They are consistently neglected in assessments of yields, productivity, and market value. In addition, the off-farm values and public benefits of agricultural biodiversity are rarely considered in economic assessments. Not only governments but also markets fail to value the social benefits of biodiversity at a macro level. Given these factors and the common overdependency on economic tools, decisionmakers regrettably have little incentive to take these agrobiodiversity benefits (and losses) into account.

In addition to market valuation failures, additional market forces contribute to agrobiodiversity losses. The expansion of global markets and trade liberalization trends of the late 20th century tend to have a homogenizing effect on food production and consumption patterns, and therefore on agricultural biodiversity. Although the development of export and import markets can expand the variety of foods available to consumers in certain regions, these global markets usually demand uniform specialized temperate varieties that are developed and sold by Northern-based companies, and are oriented to meet food desires of well-off industrial consumers. In turn, these market pressures obligate farmers worldwide to conform to those homogeneous demands. Furthermore, globalization of markets has been accompanied by policies for harmonization of standards, which displaces efforts for local adaptations and hinders sustainability. The role of global trade agreements, particularly GATT and TRIPs, discussed in the concluding chapter, underlie supporting such market pressures.

Property Regimes and Control of Resources

Another significant cause of agrobiodiversity erosion is tied to the inequitable control of and access to genetic resources—and to land and agricultural resources in general. Ironically, although the majority of genetic resources originate from developing tropical regions, Northern institutions and transnational companies have accrued advantages from developing such resources, and from extracting and exploiting the resources of the tropics—for breeding programs, seed banks, and other products such as medicinal plants.

Moreover, intellectual property regimes and laws have established formal patent systems and other legal means for companies and Northern institutions to maintain control of the knowledge, resources, and benefits associated with agricultural biodiversity. Legal measures influencing intellectual property rights (IPR) for plant genetic resources vary from country to country.[115] In general, however, the three main types of legislation are utility patents; plant breeders' rights (PBRs), which are intended to protect the right of the breeder of new plants to exclude others from certain uses of those varieties; and trade secrets.[116]

Many times, the local communities and traditional farmers who originally possessed and cultivated such diverse genetic resources have not been adequately compensated or recognized.[117] Since these groups are often poor and do not wield financial or political power, they have not been able to protect their rights and knowledge and conserve the resources that they deserve. Paradoxically, they often lack access to seed banks where traditional diverse genetic resources are collected and stored, and often cannot afford new technologies that have been developed through breeding and the use of diverse germplasm. In contrast, large private companies and research institutions generally wield power and capital to extract, collect, and control such genetic resources for their particular interests.

These inequities have contributed to the erosion of diversity and the exploitation or displacement of local knowledge by enterprises or institutions. Although the CBD and the FAO have established international guidelines intended to protect farmers' and indigenous peoples' rights, other conflicting laws (such as patents and breeders' rights) as well as global conventions and institutions, such as the GATT and the World Trade Organization, contradict and supercede the CBD by establishing intellectual property rights that give legal protection to "inventors" of new technologies, regardless of the origin. Moreover, at the local or national level, the formal laws and political systems affecting intellectual property rights tend to be biased in favor of the more influential economic interests, and do not ensure adequate farmer protection and fair distribution.[118]

Agricultural Extensification and Demographic Factors

The extensification of agriculture and expansion of populations are often partly responsible for agrobiodiversity losses in habitats. An estimated 80 percent of the 20 million hectares per year of deforestation is due to the conversion of forest to agricultural lands; about 15 million hectares of new agricultural land are planted each year.[119] However, national development policies generally promote this extensification and also underlie the movement of people into "frontier" areas.

Furthermore, in many tropical regions, as in the Amazon Basin, Central America, and Southeast Asia, the conversion process usually begins with logging (i.e., timber industries initially are responsible for clearing the land, and farming then follows).

Demographic factors, particularly population growth and large migrations of people to new frontier areas, are also influential causes of agrobiodiversity loss in some situations. But these factors are generally determined by other socio-economic phenomena, such as growth in income levels or education, or by development policies.[120] High fertility rates are linked to underdevelopment and poverty. Together, these factors result in rising demand for food and services and increased pressure on natural resources. Countries with higher population growth rates have usually experienced faster conversion of land to agricultural uses.[121] Yet, movements of people and disparities in the distribution of populations and in food marketing systems often influence agricultural patterns more than such population growth rates per se. In addition, the common trend of growing urbanization results in the expansion of urban populations into prime agricultural lands and natural areas.[122] To compensate for the loss of farmland near cities, agriculture has expanded into habitats. Migration and colonization patterns into frontier areas also can have adverse impacts on biodiversity, but as noted above, land use policies, colonization programs, and forest concession policies generally are the root causes for such movements.[123]

4

OVERCOMING CONFLICTS AND BUILDING SYNERGIES

> *"The understanding of the world's biological resources is critical to the maintenance and future balance of our food supply. Conserving and sustaining the Earth's natural resources must remain a high priority. The role agriculture plays will be a key component of a successful effort to preserve these resources."*
>
> —Howard Buffett, CEO, GSI Group

Building synergies between agriculture and biodiversity while ensuring sustainable resource use is both a major challenge and an urgent imperative. Losses and costs are multiplying rapidly. The Global Convention on Biological Diversity has established clear mandates for countries and institutions to implement actions for the conservation, enhancement, and sustainable use of agrobiodiversity.

Meeting this challenge requires major changes in practices, paradigms, policies, and commitments by governments and institutions. It also requires use of an ecosystems approach, which means integrating plants, soils, water, animals, and other resources in a holistic perspective, extending beyond the level of genetic resources. In addition, it implies a merging of agricultural, ecological, and food security factors. Members of both the environmental and agricultural communities must work together to resolve contra-

dictions and cultivate mutual interests into common achievements. A host of past and current experiences provide practical lessons and promising opportunities.

This chapter focuses on: (a) principles and "best practices" to sustainably use and enhance agrobiodiversity in farming systems; (b) participatory approaches that support local innovation and farmer knowledge, (c) strategies to merge agriculture and habitat diversity, (d) *in situ* community-based conservation, and (e) policies and institutional changes to confront the underlying causes of agrobiodiversity losses. These strategies offer "win-win" opportunities for change.

RECOMMENDED PRINCIPLES AND PARADIGM SHIFTS

The following basic principles are urgently needed to enhance and conserve agrobiodiversity. They are derived from both practical experiences and scientific research that verify the benefits of integrating biodiversity into agricultural development. They also serve as guidelines to achieve the goal of sustainable human development. These principles can be applied in all scales of production.

→ *Support sustainable ecological agriculture, which includes the goals of food security, social equity and health, economic productivity, and ecological integrity, as a framework for enhancing agrobiodiversity.*

→ *Develop an ecosystems approach, using agroecology as a guiding scientific paradigm, to support and validate the sustainable use and enhancement of agrobiodiversity at all levels.*

→ *Empower farmers and communities to protect their rights to resources, support their knowledge and cultural diversity, and ensure their participation in decisionmaking and conservation.*

→ *Adapt agricultural practices and land use to local agroecological and socioeconomic conditions, adjusted to local diverse needs and aspirations, and building upon local successful experiences.*

→ *Conserve and regenerate plant and animal genetic resources and ecosystem services using agroecological and socially beneficial methods for sustainable intensification and biodiversity enhancement.*

→ *Develop policies and institutional changes that support agrobiodiversity, ensure food security, and protect farmers' rights and eliminate policies that promote uniform monocultural systems.*

→ *Uphold and implement agrobiodiversity provisions of the Convention on Biological Diversity, as well as the mandates of the World Food Summit.*

These principles also require a significant transformation in the prevailing *paradigm* of conventional agriculture, which underlies many of these predicaments. The adoption of an ecosystems approach with a scientific basis of agroecology is an essential element of this paradigm shift. An ecosystems approach upholds biological complexity; synergism among plants, animals, and nutrients; holism, adaptability, and environmental stewardship; resource conservation; and nutrient regeneration. These concepts differ widely from the typical elements of the conventional industrial model, which features uniformity, specialization, narrowing and simplification of farming systems, and resource exploitation.

Making fundamental changes in paradigms also implies reforming strategies for research, plant breeding, and technology development and changing policies, institutions, and social systems. Research approaches need to change from an emphasis on uniformity and monocultures to an emphasis on conserving and enhancing diversity and promoting agroecological management. Likewise, intercropping, crop rotations, and frequent use of wild varieties and landraces need to be integrated into the basic research paradigms and technology development. The value of "rescuing" genetic material from wild relatives has been demonstrated with many major crops including wheat, rice, maize, sugar cane, potato, peanuts, and cotton. They provide important lessons on how diversity can be sustainably used.[124, 125]

Equally important and urgently needed to achieve these changes is a significant shift in research methodologies along with a concerted effort to improve institutional relations for technology transfer. The efforts should embrace methodological diversity, with an emphasis on using participatory methods with farmers, and upholding local knowledge and cultural diversity that is linked to biodiversity. More investment is needed in participatory approaches to research and development to complement and enhance more conventional methods. At the same time, institutional partnerships between research and development agencies, NGOs, and farmers prove to be advantageous. Responsiveness to local needs and social and cultural conditions is also vital in developing methods to support agrobiodiversity.

This paradigm shift in both agricultural and institutional or social approaches is challenging, given the economic interests and widespread institutional structures that support the conventional model. Yet, such a transformation is essential and will pay off as a foundation for implementing truly effective practices.

BEST PRACTICES THAT ENHANCE AGROBIODIVERSITY

Many existing practices that are effective and economical for farmers engaged in all scales of production also enhance agrobiodiversity. Such agrobiodiversity friendly practices are undertaken on several levels: diversification of ecosystems and crops (within broad landscapes or regions); integrating diverse types of crops, trees, and/or animals in a given system (e.g., intercropping and crop rotations, and agroforestry); mixing different plant varieties and/or breeds in farming systems; enhancing diversity in soil organisms and in insects, weeds, and fungi; conserving diversity in "natural" flora and fauna; and the conservation of diverse genetic resources. *(See Box 6.)* These practices fit within a broader agroecological approach. Similarly, biodiversity conservation methods are important features in "regenerative" or "organic" farming—which refer to ecologically based agricul-

ture approaches that virtually eliminate synthetic chemicals.

Among the important benefits of agrobiodiversity friendly practices are: improvement of nutrient cycles and soil quality;[126] increases in the sustainability and stability of systems; enhancement of productivity and contribution to food security; adding economic value by producing more diverse products; adding value to the nutrition and health status of consumers; and alleviation of pressures on habitats.[127] The particular practices selected to conserve and enhance diversity in a given farm or region logically need to be adapted to local biophysical and socioeconomic conditions. Following an agroecological approach, there are no predetermined recipes nor prescribed technology packages for a given crop or farm. Rather, producers flexibly adjust the methods to their problems and resources.

Diversity in Cropping Systems and Organic Production
Farmers in many parts of the world recognize the value of diversity in farming systems for food security, risk reduction, and productivity improvements. For example, an estimated 70 to 90 percent of beans, and 60 percent of maize in South America are intercropped with other crops. *(See also Box 7.)* Intercropping, crop rotation, and cover crops (e.g., grasses or legumes planted between rows) are used throughout the world in both traditional and modern systems. They have multiple benefits for pest/soil management, and for increasing soil fertility, improving farm income.[128]

Biodiversity enhancement practices are also commonly used in organic farming. Although organic regulations do not usually require biodiversity enhancement methods, organic producers and businesses usually acknowledge the benefits of diversity and of product diversifica-

BOX 6.

AGROECOLOGICAL AND ORGANIC APPROACHES THAT ENHANCE AGROBIODIVERSITY[a]

- Crop Diversification and Diversity Enhancement
  —temporal (crop rotation, sequences),
  —spatial (polycultures, agroforestry, crop/ livestock systems, intercropping),
  —genetic (multiple species/varieties, multilines, interspecies),
  —regional (i.e., variation in ecosystems in watersheds and ecozones).

- Recycling and Conservation of Soil Nutrients and Organic Matter
  —plant biomass (green manures, crop residues, and mulch for diverse soil nutrients),
  —animal biomass (manure or dung, urine, etc.),
  —reuse of nutrients and resources internal and external to the farm (e.g., tree litter),
  —integration of diverse plants or organisms (vermiculture or cover crops, mainly legumes),
  —strips of vegetation for soil erosion prevention (also adds to soil fertility).

- Ecologically-Based Integrated Pest and Disease Management
  —natural biological control (enhancing natural control agents),
  —imported biological control methods (e.g., adding natural enemies or botanical products),
  —diverse cropping or soil management methods to enhance natural fauna,
  —enhancing use of habitats and species in habitats.

- In Situ Germplasm Conservation[b]
  —community-based seed banks,
  —farmer-based traditional seed conservation and farmer breeding methods,
  —reserves of indigenous plants (including trees) managed by local people.

- Conserving Beneficial Fauna and Flora[c]
  —planting habitat strips within farms or buffer strips around farms,
  —use of trap crops for pest management and cover crops for soil enrichment and fertility,
  —planting or conserving trees or other vegetation around and/or in farms.

Notes

a. Many of the practices listed above are considered important in organic agriculture, as well as being central in agroecology more generally. (See text for further explanation.)

b. Plant and animal species, landraces, and adapted germplasm.

c. For multiple uses and for enhancing natural enemies.

Source: Adapted from Altieri, M. 1991. "Traditional Farming in Latin America." *The Ecologist* 21(2): 93; UNDP (United Nations Development Programme). 1995. "Agroecology: Creating the Synergism for a Sustainable Agriculture." *UNDP Guidebook Series.* New York: UNDP; and Thrupp, L.A., ed. 1996. *New Partnerships for Sustainable Agriculture.* Washington, D.C.: World Resources Institute.

BOX 7. **CROP DIVERSITY AS A KEY TO PRODUCTIVITY AND ORGANIC SERVICES**

India

Ladakh is a Tibetan, Buddhist region in the north Indian states of Jammu and Kashmir with rugged elevations of 3,500 meters and more, poor soil conditions, less than 10 centimeters of rainfall per annum, temperatures of −40°C, and very brief growing seasons. And yet, the Ladakis have developed a diversity of crops, animals, agricultural, and social practices that allows them to survive in these harsh conditions.

A key feature of their success is that they have selected and planted suitable crops in particular niches. Barley is grown everywhere except in the highest villages, while apricots, apples, and potatoes are grown in the lower valleys. Peas are harvested by hand with the nitrogen-rich nodules left in the soil. Fields are irrigated with glacial runoff and enriched with human night soil. In this diversified setting, many barley plants have more grains per stalk than most European varieties, and cereal yields average about 10 metric tons per hectare compared to yields of 1 ton per hectare in India and Africa, 2.2 tons per hectare in North America, and 1.5 in the former Soviet Union. The Ladakis also raise a variety of animals for dairy products, meat, wool, and transport. Strong social cooperation also enables the Ladakis to use these methods successfully.[1]

Mexico

Del Cabo, a producers' cooperative in Mexico that began experimenting with organic farming practices in the mid-1980s, is a successful commercial producer of a wide variety of organic crops for export throughout the year. The family farms involved are small (0.5 hectare to 5 hectares) and the climate is optimal for vegetable and herb pro-

duction. Plant diversity and soil improvement are important aspects of the group's pest and fertility management.

All farms include mixed production of winter vegetables and tropical fruits for export, with corn, sorghum, and beans for animal feed and for local consumption. Vegetables include tomatoes, cucumbers, summer squash, eggplant, onions, sweet potatoes, peas, okra, garlic, and winter squash. The corn and sorghum are also used as habitat for beneficial organisms and for windbreaks. The farmers also use green manure cover crops and crop rotation with planned fallow periods to prevent pests and diseases.

Typical returns after packing, freight, and soil and pest control ranged from US$3,000 to US$5,000 per hectare during the 1990–91 growing season. Several operational costs (such as harvesting) are higher than in conventional production, but the higher price for organic products enables Del Cabo to gain greater returns. The co-op's sophisticated skills in marketing and coordination have contributed to the success of the operation with many benefits for the community. Some problems related to weather, occasional pest infestations, and equipment access still exist, but this effort is a remarkable example of commercially successful diversified organic production.[2]

Notes

1. Abstract by Rita Banerji from Norberg-Hodge, H., J. Page, and P. Goering. 1991. "Agriculture: Global Trends and Ladakh's Future." *The Ecologist* (21)2.
2. UNDP (United Nations Development Programme). 1992. *Benefits of Diversity*. New York: UNDP. pp. 120–24.

tion as a way to improve the production system and food products, reduce risks, and increase economic returns. Based on the use of such biological approaches, the organic sector is spreading with increasing success around the world. During the 1990s, the annual growth rate of the value of the organic market averaged about 20 percent globally and about 25 percent per annum in the United States.[129]

Integrated Pest and Disease Management

Diversity is a key feature in ecological integrated pest management (IPM). Effective practices for insect management include the following:[130]

- multiple cropping and/or crop rotations, used to prevent buildup of pests;

- intercropped plants that house predators of insect pests or act as alternative host plants for pests; e.g., in Tlaxcala, Mexico, farmers grow *lupinus* plants in their corn to attract the scarab beetles, and thus protect corn; in California, cover crops are used in vineyards and orchards for similar purposes;

- using certain plants as natural pesticides; for example, in Ecuador, castor leaves that contain a paralyzing agent are used to control the tenebronid beetle;

- using weeds to repel insects; for example, in Colombia, grassweeds are grown around bean fields to repel leafhoppers, and in Chile, a shrub, *Cestrum parqui,* is used to repel beetles in potatoes;

- integrating biocontrol agents including parasites, animals, and fish that consume insect pests *(see Box 8);* and

- elimination or reduction of pesticide use to avoid adverse agroecological effects on the insect diversity in agroecosystems.

Integrated disease management practices using agrobiodiversity include the following examples:

- mixed crop stands that slow down the spread of diseases by altering the microenvironment; for example, in Central America, cowpeas grown with maize are less susceptible to the fungus *Ascochyta phaselolorum,* and to the cowpea mosaic virus; and

- use of nonhost plants that are used as a "decoy" crop to attract fungus (or nematodes) and then destroyed before the disease/pest spreads.

Soil Health

Practices for improving soil health, soil fertility and nutrient cycling also make use of and enhance agrobiodiversity. *(See Box 9.)* The health and richness of diverse soil organisms are essential as a basis for fertile, balanced, productive farming systems and healthy food products. The value of soil biodiversity has been largely overlooked, but recent studies have shown that it contributes to the overall health of plants and to soil structure and water retention as well as maintaining organic material in soils. Examples include:[131]

- compost from crop residue, tree litter, and other plant and organic residues;

- intercropping and cover crops, particularly legumes, which add nutrients and fix nitrogen;

- use of some weeds that serve as a nutrient pump to bring nutrients to the surface;

- mulch and green manures (through collection and spread of crop residues, litter from surrounding areas, and other organic materials;

- integration of earthworms (vermiculture) or other beneficial organisms and biota in soil

BOX 8.

BIODIVERSITY IN ECOLOGICAL INTEGRATED PEST MANAGEMENT

Several integrated pest management (IPM) programs in Asia have resulted in remarkable reductions and in some cases elimination of pesticide use while increasing yields in rice farming. Building in agrobiodiversity—particularly using diverse beneficial insects—is a key ingredient of these IPM initiatives. For example, during the 1980s, thousands of farmers in the national IPM program of Indonesia adopted IPM methods in rice production, which included measures to enhance the diversity of insects and restore natural pest-predator interactions in rice agroecosystems. By 1992, about 100 pest observers, 3,000 extension staff, and 15,000 farmers had been trained ("farmer-field schools") to observe and understand the local ecology of the planthopper and its natural enemies. As a result, during 1987–90, the volume of pesticides used on rice fell by more than 50 percent while yields increased by about 15 percent. Farmers' incremental net profits were approximately $18 per farmer per season. The increased productivity is attributed largely to the deployment of biocontrol agents, as well as the use of pest-resistant varieties and human resources development. (Recent economic and political difficulties in Indonesia have jeopardized the project, but the initial benefits and promising outcomes were still remarkable.)

In Bangladesh, the NOPEST and INTERFISH projects also illustrate the value of this approach. Also using field schools, these programs aimed to restore the natural balance of insects and other fauna—which can eliminate pesticides and increase returns. They have also integrated fish into rice paddies, as well as planting vegetables on the dikes around the edges. Thousands of farmers involved in the program have adopted these methods and thereby gained increased rice yields and new nutritional sources while also avoiding hazardous chemical use.

Farmers in the pilot NOPEST program achieved an 11-percent increase in rice production, while eliminating pesticides. By the end of the first season, 76 percent of the farmers in the project had stopped using pesticides and the majority supplemented their nutrition through diversification.

Additional examples of effective IPM programs that incorporate diverse practices are found throughout the world. The experience of Cuba is particularly remarkable; the government has established a national program that focuses on the massive production and implementation of biologically based pest control, using a large variety of biocontrols, natural predators, and enemies. An estimated 56 percent of the cropland was treated with biological controls. An important feature of the pest control programs is the hundreds of centers for reproduction of biocontrol agents, where biocontrol agents are produced at a decentralized level. Cuba is also making a strong push to diversify away from the reliance on monocultures, which also helps to reduce risks of pests.

When the International Potato Center and several NGOs introduced IPM methods in potato farming in Peru, the rate of Andean potato weevil infestation dropped from 31 percent to 10 percent of harvested potatoes in one community, and from 50 percent to 15 percent in another. Estimated net benefits were $154 per hectare. Sticky traps for leaf miner flies, along with other IPM methods, cut production costs and increased potato yields as well; estimated benefits were $162 per hectare.

Source: Thrupp, L.A., ed. 1996. *New Partnerships for Sustainable Agriculture.* Washington, D.C.: World Resources Institute. Information on Indonesia is from World Bank. 1996. "Integrated Pest Management: Strategy and Options for Promoting Effective Implementation." Draft document. Washington, D.C.: The World Bank.

BOX 9. PAYOFFS FROM DIVERSITY IN SOIL REGENERATION METHODS

Senegal

The Senegal Regenerative Agriculture Center (SRARC) is working with the Rodale Institute to promote sustainable agriculture based on soil regeneration for small-scale farmers who had been experiencing problems from soil degradation. The primary cropping system is a millet-groundnut rotation; in addition, legumes are intercropped with cereals. The use of compost has been revived and supplemented as a key technique for soil recuperation. Cows, goats, and sheep are usually kept by each household and their manure is collected for composting. Plant residues are also mixed with the compost. This method of compost from diverse sources in an integrated model has produced significant increases in organic content, fertility, and improvement of soil moisture and nutrient retention. The project is operating in 11 villages and local farmers have had a key role in developing and evaluating the technologies. Results show that farmers can obtain an increase in millet grain of more than 400 kilograms per hectare if they put on at least 2 tons of compost. Similar yield increases were achieved with chemical fertilizers, but the cost-benefit ratio and longer-term effects are important considerations. In this sense, the integrated soil management approach has restored lasting nutrients and

improved conservation of resources needed for sustainable production.

Tanzania

In 1980, a Soil Erosion Control and Agroforestry Project (SECAP) was begun in the Lushoto district. The project began with the planting of perennial Guatemala grass along contours to alleviate soil erosion and promote soil regeneration. When this did not curb erosion, the project expanded to include diversification of the contour strips, using trees, shrubs, and creeping legumes (called "macrocontours"). The combination of these methods reduced erosion by an average of 25 percent and improved soil health. The tree species are also valuable for fodder. Total yields per hectare increased by 64 percent for areas with grass strips and 87 percent for areas with macrocontours. Based on farm trials, gross marginal returns were 74 percent higher in macrocontour systems compared to conventional approaches. The fodder production allows farmers to have more than one dairy cow per average farm of 2.5 hectares as well.

Source: UNDP (United Nations Development Programme). 1992. *Benefits of Diversity: An Incentive Toward Sustainable Agriculture.* UNDP, New York. pp. 98–102 and 108–111.

to enhance fertility, organic matter, and nutrient recycling;[132] and

- elimination or minimization of agrochemicals (especially toxic nematicides) that destroy diverse soil biota, organic material, and valuable organisms.

Agroforestry

Agroforestry also illustrates the high value of agrobiodiversity.[133] The integration of trees in farming systems provides the following advantages:

- incorporates a remarkable degree of plant and animal diversity, combining conservation and use of natural resources. (In Indonesia, for example, small-holder "jungle rubber" gardens incorporate numerous tree species, providing fruits and/or timber.)

- provides a highly efficient and intensive use of resources. (In West Sumatra, for example, agroforestry gardens occupy 50 to 85 percent of the total agricultural land.)

- shelter hundreds of plant species, constituting valuable forms of in situ conservation.[134]

Agrobiodiversity in Large-Scale Commercial Farming

A common misperception is that agrobiodiversity is only feasible in small-scale farms. In fact, experience shows that large and medium commercial farming systems also benefit from incorporating these principles and practices. The following affirmation by an agribusiness leader suggests the importance of such approaches:

Crop rotations, intercropping, cover crops, integrated pest management techniques, and green manures are among the most common and successful methods used in larger commercial systems both in the North and in the South. (See Box 10.) They represent sustainable approaches for intensification. They make better use of available resources and enhance complementarities between crops. Examples are found in tea and coffee plantations in the tropics and in organic vineyards and orchards in temperate zones. (See Figure 8.)

In many of these large-scale cases, the change from monocultural to diverse systems entails transition costs, and sometimes trade-offs or profit losses for the first two or three years. However, after the initial transition, these producers have found that agroecological changes are profitable as well as ecologically sound for

"Biodiversity is not something that farmers talk about at the local coffee shop. It is, however, already a very important component of their profession. Nature provides a whole system to make farming viable. Diverse species can be enemies of pests, can degrade residue, form soil, fix nitrogen, and pollinate crops. A productive and sustainable agricultural system depends on maintaining the integrity of this biodiversity."

—Howard Buffett, CEO, GSI Group, 1997

commercial production. When considering conversion, large producers have to assess the possible tradeoffs and economies of scale, but also can realize that changing the approach affords new valuable opportunities.

Conserving Diversity in Livestock and Aquatic Systems

Preserving or even restoring livestock and aquatic diversity is a daunting challenge; in some cases, diversity losses in this sector have been nearly irreversible. Nevertheless, conservation efforts can make a difference. Researchers have found in situ conservation approaches to be more effective than ex situ conservation efforts because animal breeds can be developed in their natural environments. When managed by communities, herders, or local organizations, such efforts are more likely to be successful. At the same time, new practices are needed to revive the use of natural mixed pastures and grasslands, rather than seeding uniform grasses, which are less healthy for livestock and are susceptible to pests and noxious weeds.

Coffee Estates in Mexico

Finaca Irlanda, located in the state of Chiapas in Mexico, is one of the oldest, organic and biodynamic coffee estates in the world. The owner, a naturalist, is committed to maintaining diversity on his farm. The farm also serves as a learning center for other farmers interested in organic coffee growing. Established in 1928, the farm's main crops are Arabica and Robusta varieties of coffee, intercropped with cardamom and cacao. The 320-hectare farm also raises dairy and beef cattle. All animal and plant waste from the farm is used to improve nutrients. Various leguminous trees (more than 40 varieties) such as *Crotalaria* and *Tephrosia* provide both shade and nitrogen. Pest control methods include maintaining crop diversity on the farms, specific pruning methods, and biological pest controls such as a wasp introduced in Mexico from Africa to control the fungi *Beauvaria bassiana*. Indigenous wild animals that are threatened with extinction, such as the puma, wild boar, pheasant, and toucan, are protected on the plantation as they would be on a reserve. Some of the animals, such as the ocelot and gray fox, are used as natural predators of pests.[1]

Tea Plantations in India

The Singampatti Group of Estates, a subdivision of the Bombay Burmah Trading Corporation Ltd, is a large tea plantation in Tamil Nadu, India, that covers 312 hectares and emphasizes integrated, organic management practices that both enhance biodiversity and increase productivity. Shade trees (e.g., *Grevilla robusta*, *Erythrina lithosperma*, and *Gliricidia sepium*) are cultivated along with commercial (spice and medicinal) bush and tree species like cinchona and cardamom. The shade trees fix nitrogen, recycle nutrients, and prevent nutrient leaching. Open areas are planted with leguminous crops, such as cowpeas, soybeans, and Guatemala grass, to control erosion and weeds. The tea bushes are planted in trenches filled with compost, prunings, and castor or neem cakes. In addition to nutritional enhancement, this technique also reduces erosion on the sloped land and enables water conservation.

Cattle are a source of dairy products and income for the farm workers, and their dung is also used to produce "gobar-gas," which supplies part of the plantation's energy needs. Since the plantation is located in the 17th Tiger Reserve of India, it houses many species of endangered animals including tigers, the lion-tailed macaque, and the Malabar squirrel.

Yields on the estate are 11 percent higher than in conventional production, with the costs of cultivation about twice that of conventional tea production. However, the market price for organic tea is about 80 percent higher than conventional tea, so the estate could potentially reap higher profits.

In the case of fisheries and aquatic resources, wasteful and destructive "modern" fishing methods must be banned and alternative strategies developed that provide careful regulation, non-exploitative traditional techniques, and a "precautionary approach" to fishery management.[135] Quotas and licenses are also needed to prevent overfishing. In addition, more substantial changes in

BOX 10. | (CONTINUED)

Unfortunately, because the market is limited in how many grades of organic tea are sold, much of the organic tea has to be sold as conventional tea thus forcing the company to face a loss. But the owners believe that the profit margin will improve after the conversion has taken place.

Labor conditions are much better in these plantations. However the greatest concern of the plantation is the loss of genetic diversity of seed tea because of the tradition of planting clones. Though the clone has built-in resistance to blister blight, the managers are concerned that the uniformity is an invitation to pests and disease.[2]

Vineyards in the United States

Large-scale commercial wine-grape plantations both in the North and the South are increasingly turning to integrated crop/pest management methods and organic production. Renowned California-based wine companies such as Fetzer, Gallo, and Mondavi have converted from conventional, chemically intensive practices to diverse cover cropping and other sustainable practices. They generally plant a mixture of cover crops between rows of vines; these include clovers, vetch, native grasses, radishes, and flowering grasses. The cover crops are then mowed and incorporated into the soil several times a year. The diverse cover crop flora activate key processes of the agroecosystem. In addition, the diverse cover crops:

- improve soil structure and water penetration, adding organic matter and roots to increase soil aeration and decrease tillage requirements;
- prevent soil erosion, partly by spreading and slowing the movement of water;
- improve soil fertility by adding organic material; and
- control insect pests by harboring beneficial insect predators and parasites. (*See Figure 8.*)

These methods have proven economical, and sometimes more profitable, as well as sustainable. Grape growers in other countries have also begun to adopt such methods.[3]

Notes

1. Abstract by Rita Banerji from UNDP (United Nations Development Programme). 1992. *Benefits of Diversity: An Incentive Towards Sustainable Agriculture.* New York: UNDP Environmental and Natural Resources Group.
2. Abstract by Rita Banerji from UNDP (United Nations Development Programme). 1992. *Benefits of Diversity: An Incentive Towards Sustainable Agriculture.* New York: UNDP Environmental and Natural Resources Group.
3. Finch, C.V. and C.W. Sharp. 1986. *Cover Crops in California Orchards and Vineyards.* Washington, D.C.: USDA Soil Conservation Service, as cited in UNDP. 1995. *Agroecology: Creating the Synergism for a Sustainable Agriculture.* New York: UNDP.

industrial development, integrated watershed management, and coastal land uses are needed to avoid the broader disruptions and pollution of both freshwater and marine resources. Political commitments and partnerships among institutions and countries are also required to implement effective changes, since the management of these common valuable resources requires joint solutions.

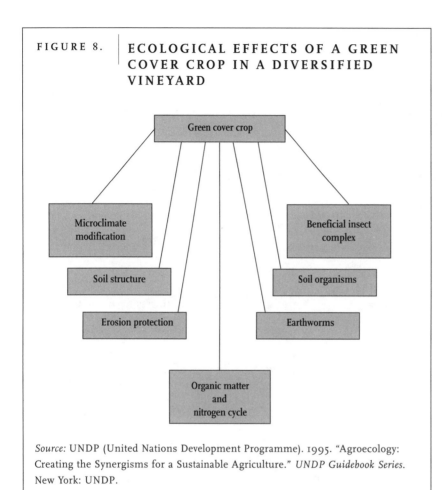

FIGURE 8. | ECOLOGICAL EFFECTS OF A GREEN COVER CROP IN A DIVERSIFIED VINEYARD

Green cover crop

Microclimate modification

Beneficial insect complex

Soil structure

Soil organisms

Erosion protection

Earthworms

Organic matter and nitrogen cycle

Source: UNDP (United Nations Development Programme). 1995. "Agroecology: Creating the Synergisms for a Sustainable Agriculture." *UNDP Guidebook Series.* New York: UNDP.

ticipation of farmers in decisionmaking as well as the full incorporation and inclusion of local farming practices and local knowledge in agricultural research and development has had beneficial outcomes.[136] *(See Boxes 11 and 12.)* In addition, the protection of farmers' and indigenous peoples' right of access to genetic resources is an essential means of empowerment and a basic need and human right.

"Scientists can improve the relevance of their research by drawing on farmers' own informal methods of experimenting with unfamiliar cultivars and practices."[137] Particular value can be gained through the full participation of women farmers because women in many areas of the world have unique knowledge about biodiversity. More work is needed to ensure women's involvement in such efforts. *(See Box 12.)*

PARTICIPATION AND EMPOWERMENT OF FARMERS

The numerous "best practices" for agrobiodiversity have proven to be beneficial, but cannot be actually implemented without active involvement of farmers and without other changes in human behaviors and social conditions, In particular, the empowerment and participation of local people with local knowledge, as well as the protection of their rights and cultural diversity are needed for implementing such practices. Many experiences have shown that the true par-

There has been much discussion about the best way to increase participatory approaches and empowerment.[138] These discussions concur that changing conventional agriculture methods obviously takes time and an investment of human resources. *(See Box 13 for general principles.)* Experiences prove that an interactive approach, with farmers taking the lead in the decisionmaking and organization of agroecological initiatives, improves the chances that agrobiodiversity enhancement efforts will succeed.

INCORPORATING INDIGENOUS PRACTICES

Mexico

In Mexico, researchers worked with local people to recreate *chinampas* (multicropped, species diverse gardens developed from reclaimed lakes), which were native to the Tabasco region of Mexico's pre-Hispanic tradition. A similar project was conducted in Veracruz that also incorporated the traditional Asiatic system of mixed farming, mixing chinampas with animal husbandry and aquaculture. Not only were these gardens rich in crop and noncrop species, but they also made more productive use of local resources and integrated plant and animal waste and cattle feed as fertilizer. Yields of these systems equalled or surpassed those of conventional systems.

Bolivia

A project called AGRUCO has introduced the wild Lupin (*Lupinus mutabilis*) into the agro-pastoral systems in the highlands of Bolivia as a replacement for fertilizers. Lupin, which has been used by the highlanders for thousands of years, is intercropped or used in rotation with other crops because of its capacity to fix up to 200 kilograms of nitrogen per hectare in the soil.

Burkina Faso

A project on soil conservation and integrated cropping in the Yatenga province was based largely on an indigenous technology of Dogon farmers in Mali, building rock bunds to prevent water runoff.

The bunds were built along contour lines and also revived an indigenous technique called "zai" in which compost is added to holes in which seeds of millet, sorghum, and peanut are planted. These crops are in a multicropping system and animals are incorporated for their manure. Yields were consistently higher than with conventional practices and ranged from 12 percent above conventional yields in 1982 to 91 percent in 1984. With the zai method yields reached 1,000 to 1,200 kilograms per hectare (kg/ha), instead of conventional yields of 700 kg/ha. Water management was enhanced and food security, a priority concern of local people, was also improved through this approach. By 1988, more than 3,500 hectares were farmed using these methods.

Source: Project summaries on Mexico and Bolivia are by Rita Banerji from Altieri, M. and L. Merrick. 1988. "Agroecology and *In Situ* Conservation of Native Crop Diversity in the Third World." In *Biodiversity*. Washington, DC: National Academy Press; and Morales, H.L. 1984. "Chinampas and Integrated Farms: Learning from the Rural Traditional Experience." In F. De Castri, G. Baker and M. Hadley, eds. *Ecology and Practice: Volume 1—Ecosystem Management*. Dublin: Tycooly. Summary of Burkina Faso is from UNDP (United Nations Development Programme). 1992. *Benefits of Diversity*. New York: UNDP, p. 95.

Similarly, it is also important to revive and restore cultural diversity, which has been repeatedly disrupted and displaced by monocultural systems and by the global inundation of uniform Western or Northern food products and technologies. The revival of cultural diversity in food systems has not only socioeconomic and cultural advantages for coping with stresses and uniting communities, it also has economic advantages for cultures that have gained income by market-

THE ROLE OF WOMEN IN DIVERSIFIED PRODUCTION AND PLANT BREEDING

Experience has shown that the full involvement, participation, and leadership of women pays off in research and development initiatives tied to agrobiodiversity. In a plant breeding project of the International Center for Agricultural Research in the Semi-Arid Tropics (CIAT) in Rwanda, scientists worked with women farmers from the earliest stages of the project to develop new varieties of beans to suit the needs of local people.

Women in Rwanda, like most in Africa, are responsible for a large majority of farming: they have valuable knowledge about preferred characteristics of varieties, seed sources, and diversity. The women were invited to experiment, manage, and evaluate trials, and to make decisions on the trial results. Through a deliberate process of participation and joint planning and learning, the scientsts and local women identified characteristics that would improve existing varieties of beans.

These experiments, undertaken both on station and on farms, resulted in stunning outcomes: The bean varieties selected and tested by women farmers over four seasons performed better than the scientists' own local mixtures in 64 percent to 89 percent of the trials. The women's selections also produced substantially more beans, with average increases in yields as high as 38 percent.

Similar experiences highlighting women's participation have proven effective in countries such as India, Ecuador, and Kenya, not just in crop breeding, but in a variety of systems and projects involving agrobiodiversity management.

Source: CGIAR (Consultative Group on International Agricultural Research). 1994. *Partners in Selection.* CGIAR, Washington, DC.

---

ing traditional diverse crops to consumers, who then begin to appreciate a wider array of diverse foods high in nutritional quality as well as cultural meaning. *(See examples in Box 11.)*

STRATEGIES TO MERGE AGRICULTURE AND HABITAT BIODIVERSITY

Biodiversity-based Intensification to Avoid Extensification

The agroecological practices summarized in the previous section are also modes of sustainable intensification; they not only increase productivity and sustainability, but can simultaneously help alleviate pressures on natural habitats and the wider resource base. For example, agro-

forestry, multiple cropping, crop rotations, integrated and organic pest and soil management, and integration of wild species can increase yields, use resources more efficiently, and minimize the need to expand farmland. In the Amazon, for instance, intensive agricultural practices (e.g., cultivation of black pepper, oranges, and vegetables) have been shown to be more profitable than extensive practices such as shifting cultivation and ranching.[139] Complex agroforestry systems are particularly beneficial because they serve biodiversity conservation by mimicking more diverse natural ecosystems and housing useful fauna and flora.

Methods involving sustainable intensification generally require more management skills and

BOX 13.   PARTICIPATORY APPROACHES IN AGROECOLOGICAL
RESEARCH AND DEVELOPMENT

- Joint problem-solving among local people, scientists, and/or extension workers (to identify challenges and options);
- Mutual listening, learning, and respect (technical and scientific personnel learn from farmers);
- Understanding social and cultural complexity as well as agroecological diversity;
- Using an inductive approach (based on local observations and data, not preconceived ideas);
- Using flexibility in selecting research and development methods, adjusting to local needs and conditions;
- Developing an interdisciplinary and holistic perspective;
- Ensuring inclusive equitable representation (gender, class, ethnicity, age) in the participatory activities;

- Building empowerment and capacities of local people in all stages of R&D; and
- Convening diverse community groups for planning, decisionmaking, and implementation.

Source: Adapted from CIDE/ACTS. 1990. *Participatory Rural Appraisal Handbook.* Washington, DC: World Resources Institute; Thrupp, L.A., B. Cabarle, and A. Zazueta. 1994. "Participatory Methods in Planning and Political Processes: Linking the Grassroots and Policies for Sustainable Development." In I. Scoones and J. Thompson, eds. 1994. *Beyond Farmer First.* London: IT Publications.

knowledge and often more labor inputs than conventional methods of uniform monocultural growth. As such, in some remote areas, these requirements may not be accessible for poor people such as settlers in newly colonized areas.[140] However, these methods often enable farmers to obtain significant yield increases per unit of land and they have broader social/ecological advantages.

Revegetation: Buffer Zones, Corridors, and Habitat Strips
Revegetation usuallly refers to planting native plants and trees in "remnant" habitat areas or forest margins that have become degraded or are unproductive for agriculture.[141] It is a means of increasing biodiversity near or within agricultural areas. Revegetation can be accomplished by *(see Figure 9):*

- Buffer strips with native plants are placed around existing remnants of natural vegetation to protect them from external impacts and/or dense vegetation in a narrow band around the edge. Trees or other forest products can be added to provide additional value.

- Corridors with strips of vegetation or trees are planted between isolated habitat remnant areas. These create a conservation "network" for fauna (particular species or groups of species or ecosystems) while also providing windbreaks and erosion control.

- Increasing remnant habitat by establishing a variety of compatible local species alongside "patches," to recreate ecosystems that are structurally and functionally similiar to existing native vegetation.[142]

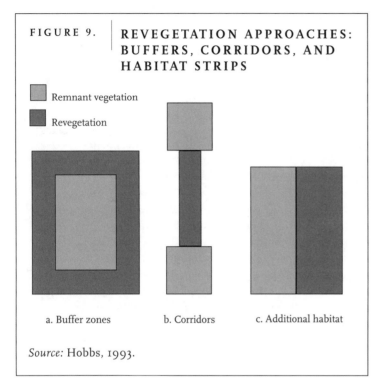

FIGURE 9. | **REVEGETATION APPROACHES: BUFFERS, CORRIDORS, AND HABITAT STRIPS**

☐ Remnant vegetation

☐ Revegetation

a. Buffer zones   b. Corridors   c. Additional habitat

Source: Hobbs, 1993.

that is potentially productive for agriculture. Specific benefits and costs need to be weighed, vis-à-vis public and private resources available. Any assessment of relative values, however, should include a full valuation of the resource-related values over the short-term and the long-term. When viewed this way, revegetation often is favorable.

Buffer-zone management, where areas of vegetation larger than just strips are involved, can be used to shelter a biological or forestry reserve from agricultural (or urbanized) lands. Buffer zones are usually placed around forest preserves to establish a semi-protected area of revegetation. They may also include sustainable farming practices, particulary agroforestry, or involve the use of nontimber forest products.

Another type of vegetation strategy is planting natural habitat strips *within* farming systems (between rows or around farm edges) to provide flora and fauna that can enhance both production and conservation.

Experiences with these techniques are relatively limited, and the regrowth process is usually quite slow, but these practices have been successfully used in various areas (such as in orchard systems in California) and they are practiced traditionally in some regions. The selection of a particular technique depends on the extent and type of degradation, the conservation goals, and the resources available.

While these habitat strategies can serve multiple goals for conservation and agricultural productivity, there may also be trade-offs. Planting a buffer strip, for example, may take up land

The Guatemalan government, for example, has employed buffer-zone management in the Peten surrounding the remaining biosphere reserve. This buffer uses regenerative farming practices, particularly agroforestry, to increase sustainable intensification and prevent the uncontrolled spread of agriculture into marginal areas. Other programs have been established in Madagascar and Peru. *(See Box 14.)* These programs, sometimes known as integrated conservation and development projects (ICDPs), can serve as an important incentive to communities to refrain from using resources in core protected areas while farming sustainably.

However, these kinds of measures in buffer zones have been difficult to develop and sometimes have been controversial because the relationships between the community needs and the park conservation interests have not been clearly defined or mutually beneficial. Local people are not always full participants and beneficiaries, and land-use restrictions often prevent communities from gaining benefits after these buffer zones are established. More attention is needed to ensure that local people are fully involved and benefit from such programs and to adequately address both conservation and development needs in these areas of biological richness.

SUPPORTING *IN SITU* AND COMMUNITY-BASED CONSERVATION INITIATIVES

Conservation of genetic resources is another essential strategy for restoring agrobiodiversity. Indigenous varieties and landraces are particularly important because they are seriously threatened. Species collected from habitats are also important sources of food, particularly for poor farmers who benefit from a diversity of nutritional sources. Farmers need to have access to such sources not only for basic livelihoods but to reduce economic risk.

This habitat diversity is also important in crop breeding and disease and pest control. A single gene imported from a barley variety in Ethiopia has been used in breeding to protect the California barley crop (worth US$160 million) from a lethal yellow dwarf virus. A wild rice gene from India helps to protect rice production in Asia from four main rice diseases.[143]

There are two main approaches to genetic resource conservation: *ex situ* (outside the source) and *in situ* (on site). *The ex situ* approach has predominated for many years, mainly through gene banks (also known as seed banks) set up by agricultural research institutions, universities, and other scientific agencies who use these large stores of diverse germplasm as a basis for breeding and research. The United States National Seed Storage Laboratory (NSSL) has the single largest seed collection in the world. It is also the center of a network of seedbanks known as the National Plant Germplasm System that contains nearly 557,000 samples of seeds, plant cuttings, tubers, and roots.[144] The system depends on supplies from the South for the majority of their acquisitions. The collections offer a broad assortment of cultivated varieties and wild and weedy species. Other major seed banks are found in the CGIAR centers and national gene banks.

Gene banks suffer from several flaws, mainly because the seed samples and germplasm are separated from agricultural production outside the ecological context where they originate, preventing the evolution of features that allow for effective adaptation.[145] Furthermore, the collections in gene banks focus on conventional crops and are located mostly in the Northern countries

where they are often inaccessible to farmers of the South. Seedbanks suffer from other deficiencies as well, such as poor infrastructure and maintenance; inadequate systems of documentation, conservation, and seed regeneration; and deficiencies of wild races.[146] *(See Box 15.)* While such banks can be important, communities cannot salvage disappearing plant species through closed collections or reserves outside natural habitats.[147]

On the other hand, *in situ* conservation efforts on farms and in communities offer an effective and valuable form of conservation. These consist of community-based seed maintenance programs and storage of indigenous and rare varieties; farmers' cultivation and management of multiple varieties and landraces in their own farms; and decentralized systems of seed selection, crossing, and exchange among farmers.[148] *(See Table 5.)* Such techniques are particularly effective in developing countries where indigenous varieties have been under threat. The seeds and germplasm are accessible and under the control of local people. Furthermore, in some areas of the Andes and in Mexico, community farming groups have also developed innovative "indigenous seed fairs," where local farmers display their conserved varieties and are publicly recognized for their conservation achievements.[149] These efforts uphold cultural diversity and local knowledge, and contribute to empowerment of those who participate. These initiatives are, unfortunately, all too rare.

The farmer-centered *in situ* approach to conservation has been endorsed by major international agreements, particularly by the Leipzig Global Plan of Action (on Plant Genetic Resources) based on a conference of participants from 158 nations, convened by the FAO in 1996. This was a very important catalyst of global support among governments, scientists, NGOs, farmers, and others for agrobiodiversity

BOX 15. CONSTRAINTS OF SEED BANKS

Seed banks constitute a significant approach to *ex situ* conservation of plant genetic resources. However, in many cases, they suffer serious inadequacies. Fewer than 1 percent of the 2 million germplasm accessions held around the world include extensive data.[1] Sixty-five percent lack even basic data on the source of the germplasm; 80 percent lack data on useful characteristics, including methods of propagation; and 95 percent lack any evaluation data such as results of germinability tests.

The U.S. National Seed Storage Laboratory (NSSL), although renowned for its extensive collection, has been sorely neglected. Estimates indicate that 90 percent of seed samples brought into the U.S. before 1950 were lost because of inadequate knowledge and lack of suitable storage facilities.[2] A major review of seed banks by the International Board of Plant Genetic Resources (IBPGR) in 1987 revealed major inadequacies in many of the international centers as well; 7 of the 17 seed banks evaluated did not meet IBPGR's registration standards.[3]

These poor conditions can contribute to genetic erosion. For example, many of the corn samples in the International Center for Maize and Wheat Improvement (CIMMYT) are jeopardized because of inadequate refrigeration, germination, and maintenance.[4] These weaknesses are partly due to a lack of funding by government agencies.

In addition, many collection and conservation programs of gene banks tend to focus only on selected crops. Crops that are considered by economists as "less lucrative" today are being lost because they are often not included in collections. Yet these crops actually have high economic and social value. Also disconcerting is the fact that most of the world's national seed banks are located in Northern countries rather than in the South even though the main sources of diversity are in the South. Many developing countries lack the capital, technical capacities, and political will for genetic resources.[5]

Notes

1. Peters, J.P. and J.T. Williams. 1984. "Towards Better Use of Genebanks with Special Reference to Information." *Genetic Resource News.* Food and Agriculture Organization of the United Nations (FAO) 60:22–32.
2. Myers, N. 1983. "A Wealth of Wild Species," quoted in Raeburn, P. 1995. *The Last Harvest.* New York: Simon and Schuster.
3. IBPGR. 1987. *Progress on the Development of the Register of Genebanks.* Rome, Italy, cited in Juma, C. 1989. *The Gene Hunters.* Princeton, New Jersey: Princeton University Press, p. 256.
4. Raeburn, P. 1995. *The Last Harvest.* New York: Simon and Schuster.
5. Raeburn, P. 1995. *The Last Harvest.* New York: Simon and Schuster.

conservation, local peoples' rights, and farmer-based efforts. The Convention on Biological Diversity and the International Undertaking on Plant Genetic Resources are additional global initiatives that establish global provisions to support this approach. While this macro political endorsement is important, more practical action and investments are urgently needed to strengthen and spread these effective approaches. Furthermore, to ensure that *in situ* conservation areas remain accessible to poor communities, and are protected from exploitation, community-based

TABLE 5 | CONSERVATION UNITS FOR *IN SITU* CONSERVATION OF GENETIC RESOURCES

Conservation Unit	Wild Populations or Near Relatives	Landraces or Traditional Breeds
Farms, home gardens, ranches		X
Indigenous reserves in communities		X
Extractive reserves	X	
Managed forests (by local people)	X	
Religious sanctuaries and shrines	X	X
Range land managed by pastoralists	X	X
"Wildland" reserves (managed by locals)	X	

Source: Adapted partly from Srivastava, J, N. Smith, and D. Forno. 1996. "Biodiversity and Agriculture: Implications for Conservation and Development." World Bank Technical Paper 321. Washington, DC: World Bank.

tenure systems need to be strengthened or modified. These kinds of initiatives can help contribute to food security and social equity as well as ecosystem functioning.

ADDRESSING ROOT PROBLEMS: POLICIES, PARADIGMS, AND PROTECTING RIGHTS

Changes to policies and institutional structures affecting agrobiodiversity are urgently needed. Practical changes at the local level are sometimes difficult, if not impossible, if such macro forces are not involved. Some groups and institutions, from the grassroots to global levels, have attempted to address these challenges. However, these efforts are not enough. More effective policy reform and implementation, as well as institutional transformation, is needed to confront the roots of problems and ensure agrobiodiversity conservation, food security, and sustainable development.

Institutions and Initiatives Affecting Agrobiodiversity Policies
Some of the most important initiatives affecting agrobiodiversity are outgrowths of local efforts by communities, NGOs, grassroots farmers' associations, and women's groups. Many are working not only on field-based practices but on policy and advocacy work, to enhance, conserve, and ensure their rights to the benefits of agrobiodiversity. Although individually, they are often small in scale, collectively, they can make a big difference.

In India in 1995, approximately 500,000 farmers carried out a peaceful demonstration for the protection of farmers' rights to plant genetic resources, opposition to IPR regulations in GATT, and the support of alternative integrated agroecosystems. The Greenbelt movement of women in Kenya is a major national movement involving thousands of women who promote planting and conservation of diverse indigenous trees on farms—through political advocacy and local-level programs. More broadly, in 1996, representatives from 120 NGOs in 50 countries gathered in Leipzig to agree on an important commitment to agrobiodiversity and farmers' rights. The renowned "People's Plan of Action" was the result of this meeting. Similarly, the Crucible Group is another international coalition of people from NGOs and other interest groups that has addressed IPR.[150] A farmer-centered international initiative on Community-Based Plant Genetic Resources is supporting indigenous groups in several countries.[151] Such initiatives are emerging in many areas; they represent important public support and future opportunities for alternative approaches, policies, and practices. Yet, they still have few resources and limited economic influence. Such grassroots initiatives and coalitions need more support and multiplication for positive actions and policy changes.

On the other hand, at the global level, several international institutions have been involved in research, development, and policy formulation that influence the use of genetic resources and biodiversity in agriculture. *(See Table 6.)* One of the influential institutions, the Consultative Group on International Agricultural Research (CGIAR), is comprised of a system of International Agricultural Research Centers (IARCs) that began in the 1950s. This system developed major plant collections, breeding, and research programs. About this time, FAO also began playing a major role in supporting plant breeding and outreach.[152] FAO has continued to play an active role in many regulatory issues and research in this field. In 1974, the International Board of Plant Genetic Resources (IBPGR) was also set up as a member of the CGIAR to focus on plant genetic resources; in 1994 the name was changed to the International Plant Genetic Resources Institute (IPGRI).[153] For many years these institutions have dedicated considerable resources to developing high yielding varieties

and technologies for the Green Revolution. More recently, however, they have increasingly developed initiatives to conserve genetic resources. These institutions have also developed policies affecting the use, conservation, and control of plant genetic resources and intellectual property rights in response to growing public concerns.[154]

IPR policies have generated great debate, discussion, and a large body of literature in recent years.[155] Major concerns revolve around deciding the appropriate roles for institutions, companies, and community groups that influence access to these resources; ensuring the fair distribution of benefits from plant genetic resources; and protecting local peoples' rights.[156]

International policies and bodies have been developed to address such legal matters. The Union for the Protection of New Varieties of Plants (UPOV), first established in the 1960s, focuses on new innovations for crops/plants and breeding rights, and has held conventions to address the regulations.[157] Another entity is the Commission on Plant Genetic Resources (CPGR), developed through the FAO, which in 1987, accepted IPR protection for breeders in exchange for recognition of farmers' rights.[158] In contrast, the International Undertaking on Plant Genetic Resources was also formed as an agreement to recognize local communities and farmers' contributions and interests in developing and conserving plant genetic resources. Such contradicting policies and mandates create confusion and require coordination to ensure the protection of the rights of farmers and indigenous peoples.

Agrobiodiversity issues have been addressed in several international conventions and agreements. As noted earlier, the Convention on Biological Diversity (CBD), established in 1992, is an important agreement to promote sustainable

TABLE 6.

TABLE 6. | MAJOR INSTITUTIONS AND CONVENTIONS INFLUENCING PLANT GENETIC RESOURCES

Acronym	Institution/Convention	Role/Influence on Plant Genetic Resources
IPGRI	International Plant Genetic Resources Institute	Coordinates world network of research centers, labs, and gene banks concerned about plant genetic resources (formerly the International Board of Plant Genetic Resources).
NSSL	National Seed Storage Laboratory	Maintains genetic materials as a base collection for the United States and for the global network of genetic resource centers.
FAO—CPRG	FAO Commission on Plant Genetic Resources	Develops guidelines and norms for intellectual property rights and policies concerning plant genetic resources.
FAO—SIDP	Seed Improvement and Development Program	Promotes participation of governments, NGOs, and industries to develop HYVs and related inputs.
WTO	World Trade Organization	Oversees and establishes international norms and model laws concerning patents and control of information on plant genetic resources.
UPOV	Union for the Protection of New Varieties of Plants	International Convention (held in 1961, 1972, 1978, 1991) that establishes regulations on plant innovations and varieties.
CBD	Biodiversity Convention (and Secretariat)	International agreement that establishes legally binding codes of conduct, guidelines, regulations.
TRIPs	Trade-Related Intellectual Property Rights	Provides minimum standards for member countries on IPR, patents and plant protection, under GATT.
GATT	General Agreement on Tariffs and Trade	International agreement concerning international trade and commerce, including provisions on IPR.

Note: Descriptions are based on interpretations of literature, not formal statements by the institutions.

use, conservation, and fair distribution of diverse biological resources globally.[159] The Convention sets out provisions that countries must implement to meet these aims, and includes mandates on agrobiodiversity. It highlights the rights of countries to have access to technologies that could assist in conservation or contribute to sustainable use of biological resources; and it calls on countries to establish measures to regulate IPR and prospecting of PGRs.[160] Recent decisions from the Conference of the Parties on the CBD, held in May 1998, in Bratislava, Slovakia, affirm strong endorsement for the implementation of actions and policies for agrobiodiversity conservation, stressing the urgency for sustainable agriculture technologies and related strategies to counteract threats. The decisions highlighted the need to support sustainable agriculture, to assess agrobiodiversity losses, and to report on the effects of trade liberalization on biodiversity. They also recommend using a precautionary approach to evaluate and control new biotechnologies (such as seeds engineered to terminate germination after one growing season) that pose risks to agrobiodiversity and food security.[161]

Another significant global agreement affecting biodiversity is the Leipzig Global Plan of Action, mentioned above. It establishes recommendations focused on conservation of plant genetic resources, including strong support of farmers' rights and local knowledge. As an effort to establish legally binding measures to implement this Plan, an International Undertaking is being negotiated among nations as well; this may be used to complement the CBD provisions.[162]

Of course the World Food Summit is another major convention that is very important to this issue. The convention itself does not include explicit details on agrobiodiversity; and people working on biodiversity rarely refer to this convention. But it establishes critical global mandates for ensuring food security for all people. The Summit also includes a commitment to sustainable agriculture practices. It is important to merge these goals and provisions for food security with environment and biodiversity aims.

Although these global conventions and political statements are important to provide a broad policy framework, they are not enough to make all of the changes needed in practice. In fact, most of the provisions are not legally binding. The written mandates must be implemented, through policy action at the national and local level, and though true political commitments to agrobiodiversity conservation and enhancement.

In addition, measures must be taken to address global conventions and policies that conflict with the CBD and erode biodiversity. For example, as noted before, the General Agreement on Tariffs and Trade (GATT) significantly influences intellectual property rights. In contrast to the CBD, it protects companies' and individuals' rights to patent protection over genetic resources.[163] Trade-related Intellectual Property Rights (TRIPS), part of GATT, conflict with CBD provisions on the farmers' rights to plant genetic resources. The World Trade Organization (WTO) establishes and enforces regulations that affect trade and rights of PGRs, as the preeminent enforcement and implementation body of GATT.[164] These international conventions and related trade liberalization policies need significant revisions, given their contradictions with the CBD, and should be aligned with the provisions of the CBD.

Policy and Institutional Changes for Agrobiodiversity and Food Security

Without significant policy transformations, it is unlikely that agrobiodiversity-enhancing practices can be widely adopted at the local level. One crucial policy change would be to eliminate the Green Revolution era subsidies and incentives that encourage uniform high-yielding varieties and agrochemicals. There is also an urgent need to reformulate trade and market policies so that they reflect ecological and social values. At the same time, national policies to support agroecological approaches must be established and rigorously implemented. This requires drafting national policies and dedicating resources that focus on biodiversity friendly methods such as agroforesty or integrated pest, crop, and soil management.

An equally important strategy is to build public participation in the formulation of agricultural, environmental, and land-use policies, and then enforce laws that protect local farmers' and communities' rights to resources. Still another urgent priority is to implement laws that will ensure ethical business practices and prevent unfair control of genetic resources and technology development by agricultural technology companies. *(See Box 16.)*

Institutional changes at all levels are also urgently needed to ensure that all institutions have the means to implement such policies. Institutions for agricultural research, development, extension, and education, as well as economic and environmental organizations, need to be retooled and reformed to support agrobiodi-

---

BOX 16. | **CRUCIAL POLICY CHANGES TO SUPPORT AGROBIODIVERSITY**

- Eliminate policies and incentives that contribute to agrobiodiversity erosion, particularly subsidies for HYVs, pesticides, and fertilizers; credit policies that require the use of such inputs and monocultures; and pricing and tax policies and extension programs that favor these inputs and uniform monocultural systems;
- Reform trade and market policies that contribute to agrobiodiversity erosion, including eliminating marketing standards that obligate farmers to produce uniform products, reform trade liberalization and market expansion policies (including GATT and TRIPS) to release market pressures that homogenize production processes, and develop market diversification strategies that promote diversity and sustainable/organic products;
- Establish national policy support for agroecological and integrated approaches to food production, ensuring political commitment, educational and institutional capacities, and incentives for

implementing agrobiodiversity enhancement policies and practices;
- Reform policies and laws affecting property and resource access rights to protect farmers and indigenous peoples' rights, enable fair opportunities, and ensure and implement fair benefit-sharing arrangements;
- Ensure public participation (particularly farmer involvement) in the development of agricultural, environmental and economic policies;
- Implement regulations affecting companies that develop and sell seeds, agrochemicals, and biotechnology to prevent unfair market domination, to ensure ethical marketing and advertising practices, and to control biotechnology development; and
- Establish flexibility and diversity in marketing standards to enable retail food stores and distributors to diversify varieties of produce, to reduce wasteful cosmetic standards for foods in markets, and to respond to consumer demands for an array of healthy foods.

versity as a critical part of sustainable agricultural development. This requires major investments and attention because present institutions operate under a very different paradigm and approach. Educating participants at all levels, from top-level decisionmakers to health services providers and professionals to field workers, in agrobiodiversity, agroecology, and farmer-oriented approaches is sorely needed. Similarly, education on the benefits of diversity in food and nutrition is valuable (and should be required) in schools and clinics as well as in agricultural universities.

Programmatic changes are also needed to enable institutions to address the many dimensions of agrobiodiversity, and to develop more integrated multidisciplinary and participatory approaches instead of specialized disciplinary nonparticipatory approaches. New partnerships among institutions are highly effective in this context. The collaboration among government agencies, NGOs, the private sector, and research organizations can combine complementary skills in joint efforts to promote effective agroecological changes. In many cases, NGOs offer more experiences and skills in agroecology and agrobiodiversity than can be found in formal public institutions and companies. The experiences and capacities of NGOs and farmer groups need to be supported and used in training for formal institutions. In addition, their efforts need to be replicated as important decentralized models for influential changes.

Such general policies and institutional reforms will help work toward sustainable agricultural production and the conservation of biological resources. Yet, more specific priorities and adjustments need to be established at regional, national, and local levels. Special attention is deserved for regions and areas where there are threats to high levels of agrobiodiversity, such as the "megacenters" of plant genetic resources.

Regional policy initiatives can be particularly important, to integrate and "mainstream" agrobiodiversity into broader economic and development policies and institutions, and to ensure cooperation among neighboring nations. Regional approaches require dialogue and agreements among multiple national interests and a variety of institutions and groups who affect and are affected by agrobiodiversity.

One such regional collaboration occurred recently in East Africa. A regional workshop was held in November 1997 among multiple stakeholders to identify priorities for policy changes for the conservation and enhancement of agricultural biodiversity. The resulting recommendations on policies and practices for that region are identified in *Box 17*. A similar Africa-wide workshop on agricultural genetic diversity and the International Undertaking on Plant Genetic Resources was convened among many African representatives in Ethiopia in April 1997. It also established significant agreements and priorities among participants about needed policies to protect the African farmers' rights to plant genetic resources. Other regions and nations could benefit by identifying similar priorities, through participatory cross-sectoral fora, and more important, by developing strong and comprehensive capacities for implementation of such recommendations.

CONCLUDING REFLECTIONS

The practices and policies included in this discussion are "win-win" solutions that can avoid conflicts and build synergies between agriculture and biodiversity. Such changes are urgently needed to overcome the serious threats from erosion of genetic resources and agrobiodiversity.

If they are *not* undertaken, the world's food supply will continue to be seriously jeopardized, and people's suffering from hunger will only

BOX 17. REGIONAL PRIORITIES FOR AGROBIODIVERSITY IN EAST AFRICA

- Diversify markets that can help promote agrobiodiversity (including policy improvements to support new markets and spread information on market trends).
- Promote attitude change to recognize the value of diverse crops and knowledge of agrobiodiversity.
- Promote/implement technology changes to conserve and enhance agrobiodiversity (including methods for conserving genetic resources, particularly indigenous seeds, with community participation, production level changes, and processing technologies).
- Ensure representation and active participation of farmers and farmers' groups in national, regional, and local forums and policy planning on agrobiodiversity; and promote partnerships among groups across sectors.
- Gather and systematize information on farmers' local knowledge of agrobiodiversity.
- Implement programs and policies that promote

farmer-based knowledge and practices of agrobiodiversity.
- "Domesticate" and internalize the CBD on farmers' rights and conservation and sustainable use of agrobiodiversity and IPR, including changing East Africa's laws.
- Depackage and delink technologies to enable choice by farmers and incorporate local varieties.
- Implement measures to mitigate potential negative effects of Green Revolution on agrobiodiversity.
- Incorporate organic measures and reduce reliance on agrochemicals.

Source: Based on a regional workshop held in November 1997 and organized by the African Centre for Technology Studies, the World Resources Institute, the International Center for Research on Agroforestry, the World Conservation Union–East Africa Regional Office, the World Wildlife Fund, and the United Nations Environment Programme.

worsen. Lessons from experience show that practices and approaches to enhance agrobiodiversity pay off for large- and small-scale farmers; they also serve the interests of food security and conservation, which benefit the broader public. If appropriate reforms are made in policies concerning intellectual property rights, they also can contribute to broad social interests. In sum, policies and actions to support agrobiodiversity at many levels are needed, and will lead to multidimensional economic and ecological gains in both the short term and the long term.

ABOUT THE AUTHOR

LORI ANN THRUPP is a Senior Associate and Director of Sustainable Agriculture in WRI's Biological Resources Program. She works primarily on sustainable agriculture, natural resource management policy, agroecology, and gender issues. Her recent publications include *New Partnerships for Sustainable Agriculture*, *Bittersweet Harvests for Global Supermarkets: Challenges in Latin America's Agricultural Export Sector*, and *The Diversity and Dynamics of Shifting Cultivation: Myths, Realities, and Policy Implications*.

NOTES

1. GRAIN. 1994. "Biodiversity in Agriculture: Some Policy Issues." *IFOAM Ecology and Farming.* January: 14.

2. This point is substantiated and illustrated in many references, such as: UNDP. 1995. *Agroecology: Creating the Synergism for a Sustainable Agriculture.* New York: United Nations Development Programme; GRAIN. 1994. "Biodiversity in Agriculture: Some Policy Issues." *IFOAM Ecology and Farming;* UNDP. 1994. *Benefits of Diversity.* New York: UNDP; Altieri, M., ed. 1987. *Agroecology: The Scientific Basis of Sustainable Agriculture.* Boulder, CO: Westview Press; Shand. H. 1997. *Human Nature: Agricultural Biodiversity and Farm-Based Food Security.* Ottawa: RAFI; Cooper, D., R. Vellve, and H. Hobbelink. 1992. *Growing Diversity.* London: IT Publications; Srivastava, J., N. Smith and D. Forno. 1996. *Biodiversity and Agriculture: Implications for Conservation and Development.* World Bank Paper 321. Washington, DC.

3. This is a summarized statement of the formal definition of the World Food Summit. FAO. 1996. *Rome Declaration on World Food Security and World Food Summit Plan of Action. World Food Summit 96/3.* Rome: Food and Agriculture Organization of the United Nations. See also Thrupp, L.A. 1998. "Critical Links: Food Security and the Environment in the Greater Horn of Africa." Washington, DC: World Resources Institute.

4. FAO (Food and Agriculture Organization of the United Nations). 1997. *World Food Summit.* See also, Swaminathan Foundation. 1995. *The Role of the GEF in helping to protect Agrobiodiversity of Global Significance.* Chenai: Swaminathan Research Foundation. (Unpublished report.)

5. GRAIN. 1994. "Biodiversity in Agriculture: Some Policy Issues." *IFOAM Ecology and Farming.* January: 14.

6. Brookfield, H. and C. Padoch. 1994. "Appreciating Agrodiversity: A Look at the Dynamism and Diversity of Indigenous Farming Practices." *Environment* 36(5):7–44.

7. Wilson, E.O. 1988. "The Current State of Biological Diversity." In E.O. Wilson, ed., *Biodiversity.* Washington, DC: National Academy Press, p. 15.

8. Juma, C. 1989. *The Gene Hunters: Biotechnology and the Scramble for Seeds.* Princeton, NJ: Princeton University Press, p. 14; and FAO. 1996. *State of the World's Plant Genetic Resources for Food and Agriculture.* Rome: Food and Agriculture Organization.

9. UNEP (United Nations Environment Programme). 1995. *Global Biodiversity Assessment.* Cambridge: Cambridge University Press, p. 128, quoted in Shand, H. 1997. *Human Nature: Agricultural Biodiversity and Farm-Based Food Security.* Ottawa: Rural Advancement Foundation International.

10. Prescott-Allen, R, and C. Prescott-Allen. 1990. "How Many Plants Feed the World." *Conservation Biology.* Vol. 4(4), pp. 365–374.

11. UNEP (United Nations Environment Programme). 1995. *Global Biodiversity Assessment.* Cambridge: Cambridge University Press, p. 129; and Scherf, B.D. 1995. *World Watch List for Domestic Animal Diversity* (2nd Edition) Rome: FAO.

12. Hammond, K. and H.W. Leitch. 1995. "The FAO Global Program for the Management of Farm Animal Genetic Resources." Beltsville, MD. Cited in Shand, H. 1997. *Human Nature: Agricultural Biodiversity and Farm-Based Food Security.* Ottawa: Rural Advancement Foundation International.

13. Altieri, M. 1994. *Biodiversity and Pest Management in Agroecosystems.* New York: Haworth Press.

14. Altieri, M. 1994. *Biodiversity and Pest Management in Agroecosystems.* New York: Haworth Press; Thrupp, L.A., ed. 1996. *New Partnerships for Sustainable Agriculture.* Washington, DC: World Resources Institute.

15. Pimentel, D. 1995, quoted in Raeburn, P. 1995. *The Last Harvest.* New York: Simon and Schuster, p. 196; and see also Stork, N. and P. Eggleton. 1992. "Invertebrates as Determinants and Indicators of Soil Quality." *American Journal of Alternative Agriculture* 7(1&2): 38–47.

16. UNEP (United Nations Environment Programme). 1995. *Global Biodiversity Assessment.* Cambridge: Cambridge University Press, p. 406. See also Shand, H. 1997. *Human Nature: Agricultural Biodiversity and Farm-Based Food Security.* Ottawa: RAFI.

17. Stork, N. and P. Eggleton. 1992. "Invertebrates as Determinants and Indicators of Soil Quality." *American Journal of Alternative Agriculture* 7(1&2): 39.

18. Pimentel. D. et al. 1996. "Environmental and Economic Benefits of Biodiversity," unpublished manuscript, cited in Shand, H. 1997. *Human Nature: Agricultural Biodiversity and Farm-Based Food Security.* Ottawa: RAFI.

19. Stork, N. and P. Eggleton. 1992. "Invertebrates as Determinants and Indicators of Soil Quality." *American Journal of Alternative Agriculture* 7(1&2): 38–47.

20. WRI. 1997. *Global Climate Change Initiative.* Washington, DC: World Resources Institute. Also see Shand, H. 1997. *Human Nature: Agricultural Biodiversity and Farm-Based Food Security.* Ottawa: RAFI, p. 83.

21. Brookfield, H. 1995. "Postscript: The Population-Environment Nexus." *Global Environmental Change* 5(4): 381–93.

22. UNEP (United Nations Environment Programme). 1995. *Global Biodiversity Assessment.* Cambridge: Cambridge University Press, p. 339. Shand, H. 1997. *Human Nature: Agricultural Biodiversity and Farm-Based Food Security.* Ottawa: RAFI.

23. Lynch, O.J. and K. Talbott. 1995. *Balancing Acts: Community based Forest Management and National Law in Asia and the Pacific.* Washington, DC: World Resources Institute.

24. Sherf, B., ed. 1995. *World Watch List for Domestic Animal Diversity*. Rome: Food and Agriculture Organization.

25. UNEP (United Nations Environment Programme). 1995. *Global Biodiversity Assessment*. Cambridge: Cambridge University Press. p. 965.

26. FAO (Food and Agriculture Organization of the United Nations). 1995. *Fishery Statistics*. Cited in Shand, H. 1997. *Human Nature: Agricultural Biodiversity and Farm-Based Food Security*. Ottawa: RAFI.

27. UNEP (United Nations Environment Programme). 1995. *Global Biodiversity Assessment*. Cambridge: Cambridge University Press, p. 943. Also see Juma, C. *The Gene Hunters*. London: Zed Press.

28. For more on traditional farmers' use of diversity and innovation, see several chapters in Cooper, D., R. Vellve, and H. Hobbelink. 1992. *Growing Diversity: Genetic Resources and Local Food Security*, London: IT Publications. Also see articles in DeBoef, W. K. Amanor, K. Wellard, with A. Bebbington. 1993. *Cultivating Knowledge: Genetic diversity, farmer experimentation, and crop research*. London: IT Publications, and Shand. H. 1997. *Human Nature*. Ottawa: RAFI.

29. UNDP. 1995. "Agroecology: Creating the Synergisms for Sustainable Agriculture." New York: United Nations Development Programme, p. 7. Citing Francis, C.A., ed. 1986. *Multiple Cropping Systems*. New York: MacMillan.

30. Altieri, Miguel. 1991. "Traditional Farming in Latin America." *The Ecologist* 21(2): 93–6; also see Gliessman et al. 1981. "The Ecological Basis for the Application of Traditional Agricultural Technology in the Management of Tropical Agroecosystems." *Agroecosystems* 7:173–85.

31. Thrupp, L.A, S. Hecht, and J. Browder. 1997. *Diversity and Dynamics of Shifting Cultivation: Myths, Realities, and Policy Implications*. Washington, DC: World Resources Institute.

32. Rice, R. and R. Greenburg. 1996. *Coffee, Conservation and Commerce*. Washington, DC: Smithsonian Institute. Greenburg, R. 1994. "Phenomena, Comment and Notes." *Smithsonian* 25(8): 24–7.

33. Altieri, Miguel. 1991. "Traditional Farming in Latin America." *The Ecologist* 21(2): 93.

34. Greenburg, R. 1994. "Phenomena, Comment and Notes." *Smithsonian* 25(8): 24–7.

35. Brookfield, H and C. Padoch. 1994. "Appreciating Agrodiversity: A Look at the Dynamism and Diversity of Indigenous Farming Practices. *Environment* 36(5).

36. Brown, A.H.D. 1978. "Isozymes, Plant Population Genetic Structure, and Genetic Conservation." *Theoretical Applied Genetics* 52: 145–57, cited in Cleveland et al., 1994.

37. IIED (International Institute for Environment Development). 1995. *Hidden Harvest: The Value of Wild Resources in Agricultural Systems*. London: IIED.

38. For further information on this topic, see e.g. Chambers, Robert et al. 1986. *Farmer First: Farmer Innovation and Agricultural Research*. London: IT Publications; Thrupp, L. Ann. 1989. "Legitimizing Local Knowledge: From Displacement to Empowerment." *Agriculture and Human Values*. Sum-

mer; DeBoef, W. et al. 1993. *Cultivating Knowledge: Genetic Diversity, Farmer Experimentation, and Crop Research.* London: IT Publications; and Cooper, D. et al. 1992. *Growing Diversity: Genetic Resources and Local Food Security.* London: IT Publications.

39. For more on indigenous knowledge, see Thrupp, L.A. 1989. "Legitimizing Local Knowledge: From Displacement to Empowerment of Third World Peoples." *Agriculture and Human Values,* Summer; Chambers, R. et al. 1986. *Farmer First: Farmer Innovation and Agricultural Research.* London: IT Publications; Warren, D.K., et al., eds., 1989. *Indigenous Knowledge Systems: Implications for Agriculture and International Development. Technology and Social Change Program.* Ames, Iowa: Iowa State University.

40. Altieri, Miguel. 1991. "Traditional Farming in Latin America." *The Ecologist* 21(2): 93.

41. See ECOGEN literature (especially Dianne Rocheleau, B. Thomas Slayter, etc.). Ecology and Gender Program, Clark University, Worcester, MA.

42. For documentation on women's local knowledge, see publications from the ECOGEN (Ecology and Gender) program, Department of Geology, Clark University, Worcester, MA. Also see Abramowitz, J. and R. Nichols. 1993. "Women and Agrobiodiversity." *SID Journal on Development;* and Thrupp, L.A. 1984. "Women, Wood, and Work: In Kenya and Beyond." *Unasylva. FAO Journal of Forestry.* December.

43. Banerji, Rita, personal communication, based on field research in India, 1995.

44. Cleveland, David, D. Soleri, and S. Smith. 1994. "Do Folk Crop Varieties Have a Role in Sustainable Agriculture?" *Bioscience* 44 (11): 740–51.

45. IIED (International Institute for Environment and Development). 1995. *Hidden Harvest: The Value of Wild Resources in Agricultural Systems.* London: IIED, p. 8.

46. Cited in Juma, C. 1989. *The Gene Hunters: Biotechnology and the Scramble for Seeds.* Princeton, NJ: Princeton University Press.

47. Juma, C. 1989. *The Gene Hunters: Biotechnology and the Scramble for Seeds.* Princeton, NJ: Princeton University Press, p. 41.

48. Juma, C. 1989. *The Gene Hunters: Biotechnology and the Scramble for Seeds.* Princeton, NJ: Princeton University Press, p. 11.

49. Raeburn, P. 1995. *The Last Harvest: The Genetic Gamble that Threatens to Destroy American Agriculture.* New York: Simon and Schuster, p. 40.

50. Shand. H. 1997. *Human Nature: Agricultural Biodiversity and Farm-Based Food Security.* Ottawa: RAFI, p. 22.

51. NRC. 1993. *Managing Global Genetic Resources.* Washington, DC: National Academy Press. Also, see Cleveland et al., 1994.

52. Plotkin, Mark. 1988. "The Outlook for New Agricultural and Industrial Products from the Tropics." In E.O. Wilson, ed., *Biodiversity.* Washington, DC: National Academy Press, pp. 106–16.

53. Srivastava, J., J. Lambert, and N. Vietmeyer. 1996. *Medicinal Plants: An Expanding Role in Development.* World Bank Technical Paper 320. Washington, DC.

54. WRI/UNEP/UNDP/World Bank (World Resources Institute, United Nations Environment Programme, United Nations Development Programme, and the World Bank). 1998. "Food Insecurity: A Trend Toward Hunger." *World Resources 1998–99*, p. 155.

55. FAO (Food and Agriculture Organization of the United Nations). 1998. *Agriculture and Food Security: The Situation Today—Hunger Amid Plenty*. Available online at: http://www.fao.org/wfs/fs/e/agricult/AgSit-e.htm (June 23, 1998).

56. GREAN. 1995. "Global Research and the Environmental and Agricultural Nexus for the 21st Century." Proposal/Report by University of Florida and Cornell University, p. 21; see also World Resources Institute in collaboration with the United Nations Environment Programme, the United Nations Development Programme, and the World Bank. 1996. *World Resources Report 1996–1997*. Washington, DC: World Resources Institute, p. 229 for map.

57. GREAN. 1995. "Global Research and the Environmental and Agricultural Nexus for the 21st Century." Proposal/Report by University of Florida and Cornell University, p. 18.

58. Williams, J.T. 1988. "Identifying and Protecting the Origins of our Food Plants." In E. O. Willson, ed., *Biodiversity* Washington, DC: National Academy Press, p. 241.

59. Prescott-Allen, R. and C. Prescott-Allen. "How Many Plants Feed the World." *Conservation Biology*. 4(4): 365.

60. Wilkes, 1983, or Frankel and Soule, 1981 (cited in draft by Faeth, P., ed. 1993. Agricultural Policy and Sustainability: Case Studies from India, Chile, the Philippines and the United States. Washington, DC: WRI.)

61. FAO. 1996. *State of the World's Plant Genetic Resources for Food and Agriculture*. Rome: Food and Agriculture Organization.

62. Juma, C. 1989, cited in Cleveland, D. et al. 1994. "Do Folk Crop Varieties Have a Role in Sustainable Agriculture?" *Bioscience* 44(11): 740–51.

63. Dalrymple, D.G. and J.P. Srivastava. "Transfer of Plant Cultivars: Seeds, Sectors and Society." In: Plant Cultivars and Technology Transfer.

64. Hussein, Mian. 1994. "Regional Focus News—Bangladesh." *Ecology and Farming: Global Monitor*, IFOAM. January, p. 20.

65. Shiva, V. 1991. "The Green Revolution in the Punjab," *The Ecologist* 21(2): 57–60.

66. IFOAM. 1994. "Biodiversity: Crop Resources at Risk in Africa." *Ecology and Farming-Global Monitor*. January, p. 5.

67. Mann, R.D. "Time Running Out: The Urgent Need for Tree Planting in Africa." *The Ecologist* 20(2): 48–53.

68. Fowler, Cary and Pat Mooney. 1990. *Shattering: Food, Politics, and the Loss of Genetic Diversity*. Tucson: University of Arizona Press, p. 104.

69. Fowler, Cary and Pat Mooney. 1990. *Shattering: Food, Politics, and the Loss of Genetic Diversity*. Tucson: University of Arizona Press, p. 63.

70. Harlan, J. and Bennett, quoted in Mooney, Pat. 1979. *Seeds of the Earth: A Private or*

Public Resource? Ann Arbor: Canadian Council for International Cooperation, p. 12.

71. Vallve, Renne. 1993. "The Decline of Diversity in European Agriculture." *The Ecologist* 23(2): 64–69.

72. Vallve, Renne. 1993. "The Decline of Diversity in European Agriculture." *The Ecologist* 23(2): 64–69.

73. Hammond, K. and H.W. Leitch. 1995. " The FAO Global Program for the Management of Farm Animal Genetic Resources." Beltsville, MD, cited in Shand, H. 1997. *Human Nature.* Ottawa: RAFI. The FAO definition of endangered is populations having less than 1,000 breeding females and less than 20 breeding males. Critical populations have less than 100 breeding females and less than 5 breeding males.

74. Plucknett, D. and M.E. Horne. 1992. "Conservation of Genetic Resources." *Agriculture, Ecosystems, and the Environment.* 42: 75–92, cited in Smith, N. 1996. "The Impact of Land Use Systems on the Use and Conservation of Biodiversity." draft paper, World Bank, Washington, DC, p. 23.

75. Hall, S.J.G. and J. Ruane. 1993. "Livestock Breeds and Their Conservation: A Global Overview." *Conservation Biology* 7(4): 815–25, cited in Smith, N. 1996. p. 43.

76. Hall, S.J.G., in press, cited in *Global Biodiversity.* London: Chapman and Hall, p. 397.

77. Rege, J.E.O. 1994. "International Livestock Center Preserves Africa's Declining Wealth of Animal Biodiversity." *Diversity* 10(3): 21–5, cited in Smith, N. 1996. p. 43.

78. FAO. 1992. "Livestock Breeds in Developing World Threatened." FAO Press release, January 28. Rome: Food and Agriculture Organization.

79. UNEP (United Nations Environment Programme). 1995. *Global Biodiversity Assessment.* Cambridge: Cambridge University Press, p. 965.

80. FAO. 1995. *The State of World Fisheries and Aquaculture.* Rome: Food and Agriculture Organization.

81. Weber, P. 1992. "Net Loss: Fish, Jobs, and the Marine Environment." Worldwatch Paper 120. Washington, DC: Worldwatch Institute, p. 16.

82. Rajasekaran, B. and D.M. Warren. 1994. "Indigenous Knowledge for Socioeconomic Development and Biodiversity Conservation: The Kolli Hills." *Indigenous Knowledge and Development Monitor* 2(2): 13–17.

83. Author's interviews of farmers. 1985. Field Research in Costa Rica, Organization of Tropical Studies.

84. Shand, H. 1997. *Human Nature: Agricultural Biodiversity and Farm-Based Food Security.* Ottawa: RAFI, pp. 20–21.

85. See e.g., Murray and Swezey, 1990, and Daxl, 1988 for pesticide resistance in cotton; see Kenmore, Dilts, for pesticide resistance in rice.

86. Pimentel, D., et al. 1992. "Conserving Biological Diversity in Agricultural/Forestry Systems." *Bioscience* 42(5): 360.

87. Pimentel, D. et al. 1996. "Environmental and Economic Benefits of Biodiversity,"

unpublished manuscript, p. 24, cited in Shand, H. 1997. *Human Nature*. Ottawa: RAFI, p. 81.

88. Pimentel, D., et al. 1992. "Conserving Biological Diversity in Agricultural and Forestry Systems." *Bioscience* 42(5): 360; Stork, N. and P. Eggleton. 1992. "Invertebrates as Determinants and Indicators of Soil Quality." *American Journal of Alternative Agriculture* 7(1&2): 38–47.

89. Pagiola, Stephan. 1995. "Interactions between Agriculture and Natural Habitats." Draft paper (December), World Bank Environment Department, Washington, DC.

90. For summary, see: Smith, Nigel. 1996. "The Impact of Land Use Systems on the Use and Conservation of Biodiversity." Draft paper, World Bank, Washington, DC.

91. Swezey, S. 1985. *Breaking the Circle of Poison*. San Francisco: Institute for Food and Policy Development.

92. World Resources Institute. 1994. *World Resources Report 1994–1995*. Washington, DC: WRI.

93. Nepstad, D.C., C. Uhl, and E. Serrao. 1991. "Recuperation of a Degraded Amazonian Landscape: Forest Recovery and Agricultural Restoration." *Ambio* 20(6): 248–55.

94. World Resources Institute. 1994. *World Resources Report 1994–1995*. Washington, DC: WRI.

95. Adapted from Pagiola, S. 1995. "Interactions between Agriculture and Natural Habitats." Draft paper (December), World Bank Environment Department, Washington, DC: WRI.

96. UNEP (United Nations Environment Programme). 1995. *Global Biodiversity Assessment*. Cambridge: Cambridge University Press.

97. World Resources Institute. 1996. *World Resources Report 1996–1997*. New York: Oxford University Press; Convention on Biodiversity (CBD). 1995.

98. National Research Council. 1993. *Sustainable Agriculture and the Environment in the Humid Tropics*. Washington, DC: National Academy Press.

99. World Resources Institute. 1996. *World Resources Report 1996–1997*. New York: Oxford University Press.

100. IIED (International Institute for Environment and Development). 1995. *Hidden Harvests Project Overview. Sustainable Agriculture Program*. London: IIED.

101. Mooney, Pat. 1979. *Seeds of the Earth*. Ann Arbor: Canadian Council for International Cooperation, p. 84, quoting Wilkes, Garrison. "Native Plants and Wild Food Plants." *The Ecologist* 7(8): 314. Also see IIED. 1992. Hidden Harvests project information. London.

102. De Souza, Jose. 1993. "Intellectual Property Rights." *Proceedings of International Crop Science Conference*. Also see Mooney, Pat. 1979. *Seeds of the Earth: A Public or Private Resource?* Ann Arbor: CCIC.

103. Darymple, D. 1986. *Development and Spread of High-Yielding Wheat Varieties in Developing Countries*. Washington, DC: USAID, p. 86 (cited in *World Resources Report 1992–1993*.)

104. See, for example, Thrupp, L.A. 1990. "Inappropriate Incentives for Pesticide Use: Agricultural Credit Requirements in Developing Countries." *Agriculture and Human Values*. Summer–Fall: 62–9.

105. Examples in Painter, M. and W. Durham, eds. 1990. *The Social Causes of Environmental Destruction in Latin America*. Ann Arbor: University of Michigan Press; also Dahlberg, K. 1990. "The Industrial Model and its Impacts on Small Farmers." In: M. Altieri and S.B. Hecht, eds., *Agroecology and Small Farm Development*. Boca Raton, FL: CRC Press; Shiva, Vandana. 1991. "The Green Revolution in the Punjab." *The Ecologist* 21(2): 57–60.

106. Repetto, R. and M. Gilles. 1988. *Public Policies and the Misuse of Forest Resources*. Cambridge University Press: New York; Hecht, S., L.A. Thrupp and J. Browder. 1996. "The Diversity and Dynamics of Shifting Cultivation: Myths, Realities and Human Dimensions." Washington, DC: World Resources Institute.

107. See, for example, Painter, M. and W. Durham, eds. 1990. *The Social Causes of Environmental Destruction in Latin America*. Ann Arbor: University of Michigan Press; Hecht, S., L.A. Thrupp, and J. Browder. 1996. "Diversity and Dynamics of Shifting Cultivation." Washington, DC: World Resources Institute. Hecht, S and A. Cockburn. 1993. *Fate of the Forest*. New York: Harper.

108. National Research Council. 1993. *Sustainable Agriculture and the Environment in the Humid Tropics*. Washington, DC: National Academy Press.

109. Reid, Walter. 1992. "Genetic Resources and Sustainable Agriculture: Creating Incentives for Local Innovation and Adaptation." *Biopolicy International No. 2*. Nairobi: ACTS Press.

110. Jondle, R. 1993. "Legal Protection for Plant Intellectual Property." *Proceedings of International Congress, International Crop Science I*. Madison: Crop Science Society of America.

111. Juma, Calestous. 1989. *The Gene Hunters*. Princeton, NJ: Princeton University Press, p. 81.

112. Vellvé, Renné. 1993. "The Decline of Diversity in European Agriculture." *The Ecologist*. 23(2): 64–9.

113. Juma, Calestous. 1989. *The Gene Hunters*. Princeton, NJ: Princeton University Press.

114. Pagiola, S. and J. Kellenberg. 1997. *Mainstreaming Biodiversity in Agricultural Development: Toward Good Practice*. Washington, DC: World Bank (Environment Paper 15).

115. Jondle, R. 1993. "Legal Protection for Plant Intellectual Property." *International Crop Science I*. Madison: Crop Science Society of America, p. 481–6.

116. Mooney, Pat. 1979. *Seeds of the Earth: A Public or Private Resource?* Ann Arbor: CCIC, p. 63.

117. For example, see Shiva, V. 1991. "The Green Revolution in the Punjab." *The Ecologist* 21(2). Mooney, P. 1979. *Seeds of the Earth*. Ann Arbor: CCIC.

118. De Souza, Jose. 1993. "Intellectual Property Rights." *Proceedings of International Crop Science Conference*. Also see Mooney, Pat. 1979. *Seeds of the Earth: A Public or Private Resource?* Ann Arbor: CCIC, and Juma, C.

1989. *The Gene Hunters.* Princeton, NJ: Princeton University Press.

119. Pimental, D. et al. 1986. "Deforestation: Interdependency of Fuelwood and Agriculture." *Oikos* 46: 404–412.

120. National Research Council. 1993. *Sustainable Agriculture and the Environment in the Humid Tropics.* Washington, DC: National Academy Press; World Resources Institute. 1994. *World Resources Report 1994–1995.* New York: Oxford University Press.

121. National Research Council. 1993. *Sustainable Agriculture and the Environment in the Humid Tropics.* Washington, DC: National Academy Press.

122. World Resources Institute. 1996. *World Resources Report 1996–1997.* New York: Oxford University Press.

123. National Research Council. 1993. *Sustainable Agriculture and the Environment in the Humid Tropics.* Washington, DC: National Academy Press.

124. Timothy, D., P. Harvey, and C. Dowswell. 1988. *Development and Spread of Improved Maize Varieties and Hybrids in Developing Countries.* Washington, DC: USAID, p. 55, cited in *World Resources Report 1992–1993,* p. 132.

125. Timothy, D., P. Harvey, and C. Dowswell. 1988. *Development and Spread of Improved Maize Varieties and Hybrids in Developing Countries.* Washington, DC: USAID, p. 51, cited in *World Resources Report 1992–1993,* p. 132.

126. Pimentel, David et al. 1992. "Conserving Biological Diversity in Agricultural/

Forestry Systems." *Bioscience* 42 (5): 354–62.

127. Altieri, Miguel. 1987. *Agroecology: The Scientific Basis of Sustainable Agriculture.* Boulder: Westview Press; UNDP. 1992. *Benefits of Diversity* New York: United Nations Development Programme; and UNDP. 1995. *Agroecology: Creating the Synergism for a Sustainable Agriculture.*

128. UNDP. 1995. *Agroecology: Creating the Synergism for a Sustainable Agriculture.* New York: United Nations Development Programme.

129. Anton Dunn, J. 1995. *Organic Food and Fiber: An Analysis of 1994 Certified Production in the United States.* USDA Agricultural Marketing Service.

130. These examples are largely from Altieri, Miguel. 1991. "Traditional Farming in Latin America," *The Ecologist* 21(2): 93–6; and Altieri, Miguel. 1987. *Agroecology: The Scientific Basis of Alternative Agriculture.* Boulder: Westview Press.

131. These examples are mainly from Altieri, Miguel. 1991. "Traditional Farming in Latin America." *The Ecologist* 21(2): 93–6; See also Pimentel, D. et al. 1992; and Brookfield, H. and C. Padoch. 1993. "Appreciating Agrodiversity: A Look at the Dynamism and Diversity of Indigenous Farming Practices." *Environment* 36(5): 7–20.

132. Lee, K.E. 1990. "The Diversity of Soil Organisms." In D.K. Hawksworth, ed. *The Biodiversity of Microorganisms and Invertebrates: Its Role in Sustainable Agriculture.* London: CAB International.

133. Michon, G. and H. de Foresta. 1990. "Complex Agroforestry Systems and the Conser-

vation of Biological Diversity." In *Harmony with Nature, Proceedings of International Conference on Tropical Biodiversity.* Kuala Lumpur, Malaysia: SEAMEO-BIOTROP.

134. Michon, G. and H. de Foresta. 1990. "Complex Agroforestry Systems and the Conservation of Biological Diversity." In *Harmony with Nature, Proceedings of International Conference on Tropical Biodiversity.* Kuala Lumpur, Malaysia: SEAMEO-BIOTROP. For further examples and benefits of agroforestry and related traditional polycultures, see chapters in Cooper, D. et al., eds. 1992. *Growing Diversity: Genetic Resources and Local Food Security.* London: IT Publications.

135. For more information on these kinds of strategies, see, for example, Shand, H. 1997. *Human Nature.* Ottawa: RAFI. And Weber, P. 1992. "Net Catch." Washington, DC: Worldwatch Institute.

136. Rajasekaran, B. et al. 1993. "A Framework for Incorporating Indigenous Knowledge Systems into Agricultural Extension." *Indigenous Knowledge and Development Monitor* 1(3): 21–4. For further relevant information, see examples in Cooper, D. et al. 1992. *Growing Diversity.* London: IT Publications. DeBoef et al. 1993. *Cultivating Knowledge.* London: IT Publications. Also Chambers, R. et al. 1986. *Farmer First: Farmer Innovation and Agricultural Research.* London: IT Publications.

137. Haugerud, Angelique and M. Collinson. 1991. "Plants, Genes, and People: Improving the Relevance of Plant Breeding." *Gatekeeper Series no. 30.* London: International Institute for Environment and Development.

138. See, e.g., Thrupp, L.A., ed. 1996. *New Partnerships for Sustainable Agriculture.* Washington, DC: World Resources Institute. Chambers, R., A. Pacey, and L.A. Thrupp. 1987. *Farmer First: Farmer Innovation and Agricultural Research.* London: IT Publications; Scoones, I. and J. Thompson. 1995. *Beyond Farmer First.* London: IT Publications; Alcorn J. 1989. "Making Use of Traditional Farmers Knowledge." In *Common Futures, Proceedings of an International Forum on Sustainable Development.* Toronto: Pollution Probe; Also see IIED. 1988–1995. "RRA Notes: Issues 1-present." and "PRA Notes Series." Sustainable Agriculture Program, London.

139. Toniolo and Uhl, 1995, cited in Pagiola, S. 1995. "Interactions between Agriculture and Natural Habitats." Draft paper, World Bank Environment Department, Washington, DC.

140. Pagiola, S. 1995. "Interactions between Agriculture and Natural Habitats." Draft paper, World Bank Environment Department, Washington, DC.

141. Hobbs, R. 1993. "Can Revegetation Assist in the Conservation of Biodiversity in Agricultural Areas?" *Pacific Conservation Biology* 1: 29–38.

142. Hobbs, R. 1993. "Can Revegetation Assist in the Conservation of Biodiversity in Agricultural Areas?" *Pacific Conservation Biology* 1: 29–38.

143. These examples are mostly from Plotkin, Mark. 1988. "The Outlook for New Agricultural and Industrial Products from the Tropics." In E.O. Wilson, ed. *Biodiversity.* Washington, DC: National Academy Press, pp. 106–16.

144. Raeburn, P. 1995. *The Last Harvest*. New York: Simon and Schuster, p. 59.

145. For further information, see Raeburn, P. 1995. *The Last Harvest*. New York: Simon and Schuster; Cooper, D. et al, 1992. *Growing Diversity*. London: IT Publications, and Shand. H. 1997. *Human Nature*. Ottawa: RAFI.

146. Harlan, J.R. 1976. "Genetic Resources in Wild Relatives of Plants." *Crop Science*. May–June.

147. Peters, J.P. and J.T. Williams. 1984. "Towards Better Use of Genebanks with Special Reference to Information." *Genetic Resource News* (FAO) 60:22–32.

148. Cooper, D. et al. 1993. *Growing Diversity*. London: IT Publications, and DeBoef, M. et al. 1992. *Cultivating Knowledge*. London: IT Publications. See also Montecinos, C. 1994. *Summary of International Program on Community Based Conservation of Genetic Resources*.

149. Mario Tapia, Centro Internacional de la Papa, personal communication, March 1998; and Julio Sanchez, Colegio de Postgraduados de Mexico, May 1998. See also chapter by M. Tapia in Cooper et al. 1993. *Growing Diversity*. London: IT Publications.

150. *Leipzig Commitment to Agricultural Biodiversity: Towards a People's Plan of Action*. June 1996; Crucible Group. 1994. *People, Plants and Patents*. Ottawa: IDRC.

151. Montecinos, C. 1996. Program on Community-Based Conservation of Plant Genetic Resources, program description. Santiago, Chile. CLADES; and personal communication with program members in Lima, Peru, March 1998.

152. Juma, C. 1989. *The Gene Hunters*. Princeton, NJ: Princeton University Press, p. 88.

153. Juma, C. 1989. *The Gene Hunters*. Princeton, NJ: Princeton University Press, p. 89.

154. An example from FAO of involvement in conservation is illustrated in FAO. 1996. "Draft Global Plan of Action for the Conservation and Sustainable Utilization of Plant Genetic Resources for Food and Agriculture." Rome: Food and Agriculture Organization of the United Nations.

155. See, e.g., Mooney, Pat. 1979. *Seeds of the Earth: A Public or Private Resource?* Ann Arbor: Canadian Council for International Cooperation; and Juma, C. 1989. *The Gene Hunters;* and the Crucible Group, *People, Plants, and Patents*. Ottawa: IDRC.

156. Reid, Walter, et al. 1993. *Biodiversity Prospecting*. Washington, DC: World Resources Institute, p. 23.

157. Crucible Group. 1994. *People, Plants and Patents*. Ottawa: IDRC.

158. Reid, Walter, et al. 1993. *Biodiversity Prospecting*. Washington, DC: World Resources Institute, p. 23.

159. Convention on Biodiversity, 1993. UN; See also Crucible Group. 1994. *People, Plants and Patents*. Ottawa: IDRC.

160. Reid, Walter, et al. 1993. *Biodiversity Prospecting*. Washington, DC: World Resources Institute, p. 159.

161. Draft Decisions on Agrobiodiversity, from the Conference of Parties in Bratislava, Slovakia, posted by the Secretariat for the Convention on Biological Diversity, (www.biodiv.org), May 1998.

162. Shand. H. 1997. *Human Nature*. Ottawa: RAFI. p. 29.

163. For a summary of relevant issues see Crucible Group. 1994. *People, Plants and Patents*. Ottawa: IDRC. Also see Shand, H. 1997. *Human Nature*. Ottawa: RAFI.

164. Information from discussions and personal communication from the Conference of Parties, May 1998, Bratislava, Slovakia, See also, GRAIN, 1998. Bulletin on TRIPs and the WTO affecting agrobiodiversity.

ACRONYMS AND TERMS

CBD — Convention on Biological Diversity

CGIAR — Consultative Group on International Agriculture Research

CIMMYT — Centro Internacional Para el Mejoramiento de Maiz y Trigo (International Center for Improvement of Maize and Wheat)

FAO — Food and Agriculture Organization of the United Nations

FAO-CPRG — FAO Commission on Plant Genetic Resources

FAO-SIDP — FAO Seed Improvement and Development Program

GATT — General Agreement on Tariffs and Trade

HYVs — high-yield varieties

IPGRI — International Plant Genetic Resources Institute

IPM — integrated pest management

IRRI — International Rice Research Institute

NGO — nongovernment organization

NSSL — National Seed Storage Laboratory

PGR — plant genetic resources

TRIPS — Trade-Related Intellectual Property Rights (part of GATT)

UNDP — United Nations Development Programme

UNEP — United Nations Environment Programme

USDA — United States Department of Agriculture

UPOV — Union for the Protection of Varieties

WIPO — World Intellectual Property Organization

agricultural biodiversity (agrobiodiversity)—The components of biodiversity that feed and nurture people—whether derived from the genetic resources of plants, animals, fish, or forests.

agrochemicals—Synthetic chemical inputs used in agriculture, including fertilizers and pesticides such as insecticides, herbicides, and fungicides.

agroecology—1. The application of ecological concepts and principles to the study, design, and management of agricultural systems. By integrating cultural and environmental factors into its examination of food production systems, agroecology seeks to evaluate the full effect of system inputs and outputs and to use this knowledge to improve these systems, taking into account the needs of both the ecosystem as a whole and the people within it. 2. Agroecology has been proposed as a scientific discipline that defines, classifies, and studies agricultural systems from an ecological and socioeconomic perspective (Altieri, 1987). Agroecology integrates ideas from indigenous knowledge with modern technical knowledge to arrive at environmentally and socially sensitive approaches to agriculture, focusing not only on production but also on the ecological sustainability of the productive system. While its emphasis is agricultural management in the field, its scope includes the wider social and ecological context in which the field is situated. Such an approach reflects an interdisciplinary analysis.

agroforestry—A farming system that involves the integration of woody species along with crops and/or livestock.

biodiversity (biological diversity)—The variety and variability among living organisms and the ecological complexes in which they occur.

biotechnology—Any technique that uses living organisms (or parts of organisms) to make or modify products, to change plants or animals, or to develop microorganisms for specific uses. Advocacy efforts are directed toward raising public awareness of the dangers of biotechnology and the available ecological alternatives considered safer and more efficient. Some of these urgent issues are: experimental and commercial release of genetically-engineered organisms and products and their effects on humans and the environment; bio-piracy of raw materials and indigenous knowledge and practices, and the

patenting of life forms and processes by transnational corporations; the need for a fair and objective government regulatory body on biotechnology activities; and the role of the General Agreement on Tariff and Trade (GATT) and the World Trade Organization (WTO) on these issues.

ex situ conservation—1. Maintenance or management of an organism away from its native environment. For crop germplasm, this term typically refers to maintenance in seed banks or repositories. 2. Conservation of genetic material outside of the ecosystem where it originated, most commonly in a genebank.

extensification (extensive agriculture)—An approach or method of agricultural development using large areas to raise livestock and crops with low efficiency of resource use. This method generally entails clearing of forest resources or vegetative land.

genetic resources—In a strict sense, the physical germplasm (hereditary material) that carries the genetic characteristics of life forms. In a broad sense, the germplasm plus information, funds, technologies and social and environmental systems through which germplasm is a socio-economic resource.

germplasm—The genetic material that forms the physical basis of heredity and is transmitted from one generation to the next by means of the germ cells. Also, an individual or clone representing a type, species, or culture that may be held in a repository for agronomic, historic, or other reasons.

in situ conservation—Conservation taking place 'on site' or in the original location. Until recently, it was narrowly used to describe conservation of genetic resources in their natural surrounding, normally protected from human interference. However, it is increasingly used to designate conservation on the farm, where genetic resources are developed, bred and maintained. 2. Maintenance or management of an organism within its native environment. For landraces, this term includes maintenance in traditional agricultural systems.

integrated pest management (IPM)—1. An ecologically based strategy that relies on natural mortality factors, such as natural enemies, weather, and crop management, and seeks control tactics that disrupt these factors as little as possible while enhancing their effectiveness. 2. The term (IPM) is used to identify a very wide range of pest management systems and practices. Biologically intensive IPM is an ecologically based approach to pest control that utilizes an interdisciplinary knowledge of crop/pest relationships, establishment of acceptable economic thresholds for pest populations, and field monitoring for potential problems. Management may include such practices as the use of resistant varieties; crop rotation; cultural practices; optimal use of biological control organisms; certified seed; protective seed treatments; disease-free transplants or rootstock; timeliness of crop cultivation; improved timing of pesticide applications; and removal or 'plow down' of infested plant material (Gold 1994:5).

intellectual property rights (IPR)—Laws that grant monopoly rights to those who create ideas or knowledge. These rights are intended to protect inventors against losing control of their ideas or the creations of their knowledge. There are five forms of IPR: patents, plant breeders'

rights, copyright, trademarks, and trade secrets. Intellectual property rights vary greatly from country to country.

intensification—The fuller use of land, water, and biotic resources to enhance agronomic performance. **Conventional:** Intensification achieved through greater energy inputs, chemical fertilizers and pesticides. **Sustainable/biological:** Intensification achieved through greater inputs of renewable energy sources, information and management, organic fertilizers and biological controls, and greater efficiency of resource use in a given area.

landrace—Geographically or ecologically distinctive populations of plants and animals that are highly diverse in their genetic composition.

monoculture—1. The growing of a single plant species in one area, usually the same type of crop grown year after year. 2. Production of large areas of a single crop, usually for export. Lacking the natural benefits of biodiversity such as biological control, intercropping, and crop rotation, monoculture typically relies heavily on synthetic chemical pesticides, fertilizers, and irrigation. Monoculture often threatens local and regional food security, as well as the environment and the health of farm workers, among others.

mycorrhizae—Various species of fungi that live in symbiosis with the roots of plants and are essential for soil fertility.

polyculture—The growing of more than one crop at once in the same field.

WORLD RESOURCES INSTITUTE

The World Resources Institute (WRI) is an independent center for policy research and technical assistance on global environmental and development issues. WRI's mission is to move human society to live in ways that protect Earth's environment and its capacity to provide for the needs and aspirations of current and future generations.

Because people are inspired by ideas, empowered by knowledge, and moved to change by greater understanding, the Institute provides—and helps other institutions provide—objective information and practical proposals for policy and institutional change that will foster environmentally sound, socially equitable development. WRI's particular concerns are with globally significant environmental problems and their interaction with economic development and social equity at all levels.

The Institute's current areas of work include economics, forests, biodiversity, climate change, energy, sustainable agriculture, resource and environmental information, trade, technology, national strategies for environmental and resource management, business liaison, and human health.

In all of its policy research and work with institutions, WRI tries to build bridges between ideas and action, meshing the insights of scientific research, economic and institutional analyses, and practical experience with the need for open and participatory decision-making.

WORLD RESOURCES INSTITUTE
1709 New York Avenue, N.W.
Washington, D.C. 20006, USA
http://www.wri.org/wri